Do-it-Yourself Flooring Guide

By Color Tile Supermart, Inc.

Welcome to Color Tile, your do-it-yourself home decorating supermart. We are one of the world's largest retailers of do-it-yourself products. Floorcoverings offered for sale in our nationwide chain of stores include ceramic, mosaic and quarry tile; wood parquet; no-wax vinyl tile and vinyl sheet flooring.

To complement your new floorcovering choice, Color Tile has a full selection of wallcoverings. These include paint, prepasted vinyls, mirror, cork and decorator brick tile, and ceramic, mosaic and quarry wall tile. We also have professional-quality mastics, adhesives, and tools for installation, as well as a full line of care products to keep your new flooring bright and beautiful.

Everything at Color Tile is geared to the do-it-yourselfer. Our experienced sales personnel are always ready to help with expert decorating and installation advice. With our huge, in-stock selection you'll always be able to find just what you need.

Come to Color Tile Supermart often—*where all the choices make all the difference!*

Table of Contents

ISBN 0-8249-6121-8

Published by Ideals Publishing Corporation
11315 Watertown Plank Road
Milwaukee, Wisconsin 53226

Editor/Art Director, David Schansberg

Introduction

One of the most important elements in any extensive remodeling or redecorating project is the floorcovering. There are many distinct types in thousands of styles to choose from. The goal is to select a floorcovering that has the character and quality best suited to the planned remodeling project. To reach this goal, several factors deserve careful consideration.

A floorcovering not only protects the subflooring and framework which lie beneath it, but should also serve the function of the room. Bathroom floorcoverings should be moistureproof, slip resistant, and easy to clean. Floorcoverings in high-traffic areas such as hallways, entryways, and kitchens must be durable to withstand the daily abuse they take. Low-traffic areas can be covered with less durable materials. In the bedroom, you may desire a floorcovering that deadens noise by absorbing sound or reduces heat loss through the floor.

Second, a floorcovering must serve the needs and aesthetic tastes of the homeowner. Floorcoverings come in a wide variety of patterns, textures, and colors. Quality flooring materials will last for many years, so it is important that the floorcovering enhances the overall decorative scheme of the room and of the entire home. A new floor can add a new dimension of beauty to a lifeless room by tying together other design elements such as wallcoverings and furnishings. A floorcovering should appeal to the homeowner.

Third, a floorcovering must fit the budget of the homeowner. Flooring is relatively expensive. Floorcoverings are manufactured in different ways and are made of widely varying materials in several grades. Since cost is generally dependent on quality, the key is to install the best quality you can afford. Floorcoverings made of quality materials generally prove to be more economical in the long run because they are more durable.

Another cost factor is the installation. Not very long ago, it required professional help to install most types of floorcoverings. Even if you wanted to tackle the job yourself, finding the proper installation information, tools and adhesives was difficult. As the do-it-yourself market emerged as a large viable market, flooring manufacturers developed floorcoverings that could be installed by the homeowner in a few simple steps. Now it is possible to find floorcovering, instructions, installation tools, adhesives, and maintenance products in one convenient location. It has never been easier to install floorcovering materials yourself, and the results are just as beautiful as a professionally installed floor. The cost savings of installing a floor yourself will allow you to invest in better quality materials.

Before shopping for any floorcovering—whether you prefer resilient sheet flooring or tiles, ceramic or quarry tiles, wood plank or parquet floors, or carpeting—we suggest that you carefully read this book. Selecting the most appropriate floorcovering and installing it properly requires a basic awareness of choices, careful planning, and a thorough understanding of installation techniques. The information presented here will familiarize you with each type of floorcovering available, including new products to be introduced in the near future.

If you plan to install the floorcovering yourself, the book will take you through each step in the process, from measuring the room, preparing the subfloor, and assembling the proper tools and materials, to installing a particular floorcovering.

The full-color photo section will stimulate remodeling ideas and show you some of the products available on today's market. The detailed installation instructions and step-by-step photographs will allow you to install any floorcovering. By reading this book and following the advice prescribed in its pages, you can make your home more attractive and enjoyable. Read, learn, and have fun!

Basic Floor Construction

Before the actual description of floorcovering materials is given, the base on which the floorcovering will be installed should be considered. Some floorcoverings require special underlayment preparation. Depending on when your house was constructed, types of underlayment materials and subfloor construction techniques vary. This brief chapter will familiarize you with types of subfloors and underlayments most common to residential home construction. Actual underlayment preparation for each type of floorcovering will be discussed in later chapters.

Suspended Floor Construction

There are two basic types of floors, slab and suspended. Almost all residential construction utilizes suspended flooring. Slab floors are discussed in the chapter about wood flooring. The subfloor or floor deck construction is the simplest form of residential framing work.

Sill Plate After the foundation walls – generally constructed of concrete blocks – have cured, a foundation sill plate of 2 × 8 lumber is attached to the perimeter of the foundation wall top. Sill sealer, usually a 1/2-inch-thick insulation material, is placed along the top perimeter of the foundation wall. When the sill plate is secured to the anchor bolts of the foundation wall, the sill sealer provides a tight fit over surface irregularities and prevents moisture or air infiltration. The sill plate supports the floor joists.

Floor Joists The size of floor joists depends on the distance they must span. Joists are usually 2 × 6 or 2 × 8 lumber on edge. The most common spacing interval is 16 inches on center. Joists are commonly lapped at the central support beam with the joist resting fully on the beam. Joists are toenailed into the sill plate. Double joists may be used to support bathtubs or weight-bearing partitions running parallel with the joists. To give joists greater stability, they are often bridged along the midpoint of the joist span. When all of the joists are in position, a band joist which runs the entire length of the structure is nailed to the end of each joist.

Subfloor When all joists are leveled and nailed into position, they are covered by the subfloor. It is the subfloor to which the finish floorcovering is secured. The accepted subfloor in traditional home construction has been double-floor construction which involves nailing 1 × 6 or 1 × 8 boards diagonally across the joists. In older homes the boards were nailed at right angles to the joists. In more recent structures, they are nailed at a 45-degree angle. Hardwood flooring is then nailed or screwed to the first layer. Underlayment is often used between layers.

As building products were refined and improved, single-floor construction became possible and popular. Today, most homes have this type of subfloor. Instead of narrow boards nailed to the joists, plywood sheets are fastened to the joists in a checkerboard pattern. This adds greater stability to the subfloor and requires less time and money to apply.

Underlayment When single-floor construction first became popular, plywood sheathing was used to cover the joists. Because the plywood sheathing has a rough uneven surface, underlayment had to be applied over the plywood to provide a smooth level base for resilient or soft floorcoverings. Underlayment is also needed for double-floor construction if the finish floorcovering is anything but hardwood.

There are four basic types of underlayment: sanded plywood, hardboard, particleboard and fiberboard. Particleboard has grown in popularity because of low cost and dimensional stability. Underlayment is fastened to the subfloor with staples or nails.

Several lumber manufacturers have developed an extremely smooth type of plywood that provides a finished surface that requires no underlayment and is suitable for most floorcoverings. This plywood has tongue-and-groove edges and is fastened directly to the floor joists with glue and nails. The subfloor is so structurally stable that bridging is not required between joists. The materials are significantly more expensive than conventional plywood sheathing, but you use less materials; so a saving of both time and money is realized.

Repairing the Subfloor Structure

Now that you understand basic floor construction, you can repair any subfloor problems before installing a new floorcovering. With age, a house settles on its foundation, and a floor can begin to sag or develop squeaks. It may also have been damaged by moisture, excess weight, or hard blows. Any damage to the subfloor must be repaired before installing a new floorcovering.

The most common subfloor problem is squeaking. Fortunately, most squeaks can be eliminated by one of several relatively simple repairs. More extensive

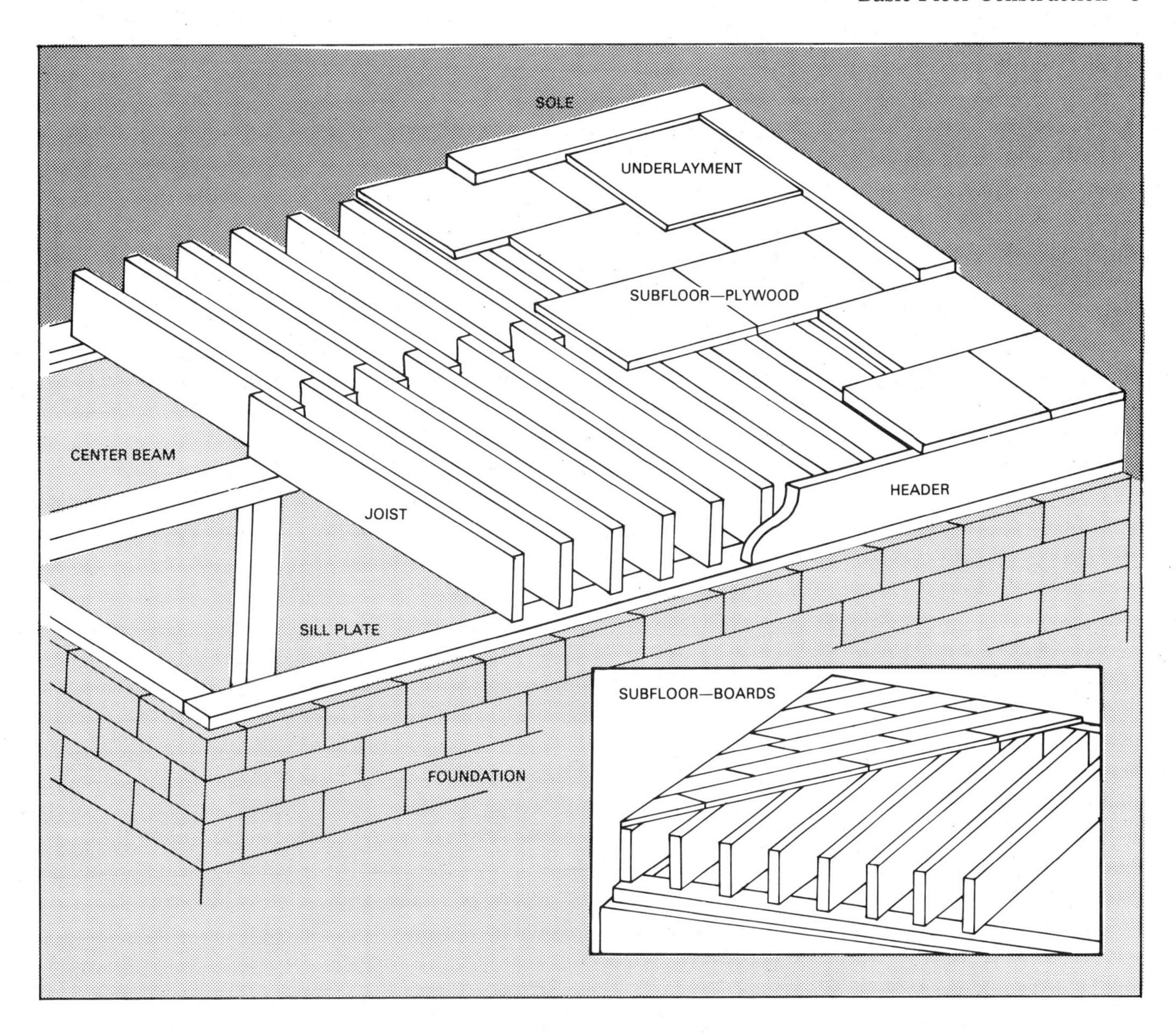

subfloor problems such as structural weaknesses, weak posts, joists rotted by moisture or insects, or a severely damaged subfloor require complicated repair procedures that should not be attempted by the novice do-it-yourselfer. Serious structural problems require professional help.

To eliminate a squeak, you must first locate it. Squeaks generally result from wood rubbing against wood. This usually occurs where subfloor boards meet the joists. As joists dry out and shrink, they can pull away from the subfloor. Improper nailing at the time of construction can also cause the subfloor to pull away from the joists. Subfloor boards can warp with age, and bridging between joists can loosen. Locating squeaks requires a little investigation and a helper.

If the squeaks are coming from the first floor of a frame house with exposed joists accessible from the basement, locating the origin of the squeak is easy. Have someone walk over squeaky areas while you observe the floor movement from below. Clearly mark the areas of greatest movement with chalk.

If the joists are not exposed, such as those between stories of a multi-level home, locating and eliminating squeaks is considerably more difficult. Squeaks in these areas must be repaired from the top surface. This is less of a problem if the old floorcovering will be removed from the subfloor prior to installing a new flooring. If this is the case, simply toenail loose boards into the joists with annular ring nails.

Several simple procedures can be used to eliminate squeaks:

1. If you detected movement between the joists and a single floor board, wooden shims can be driven into the gap to securely hold the loose board against old nailheads. Do not drive the shims in too forcefully.
2. If several consecutive subfloor boards are loose, they can all be tightened by forcing a 1 × 4 cleat against the loose boards and nailing it to the joist. Use a 2 × 4 prop to force the cleat tightly

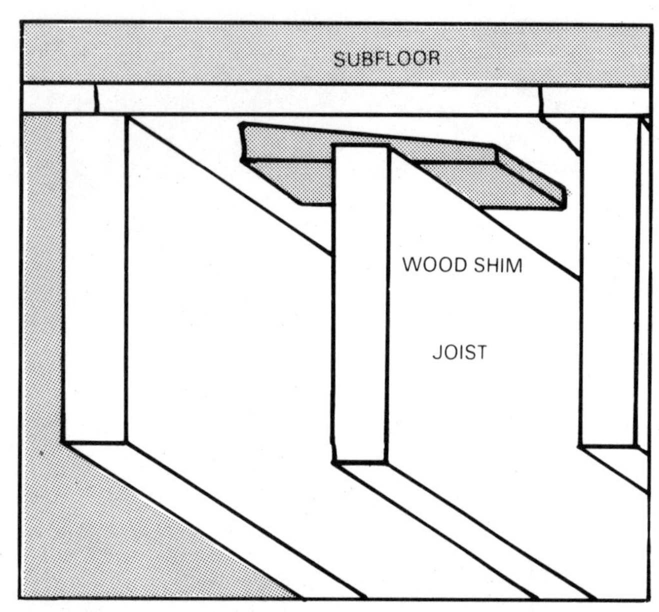

Wedging a wood shim between the joist and loose subfloor may eliminate squeaks.

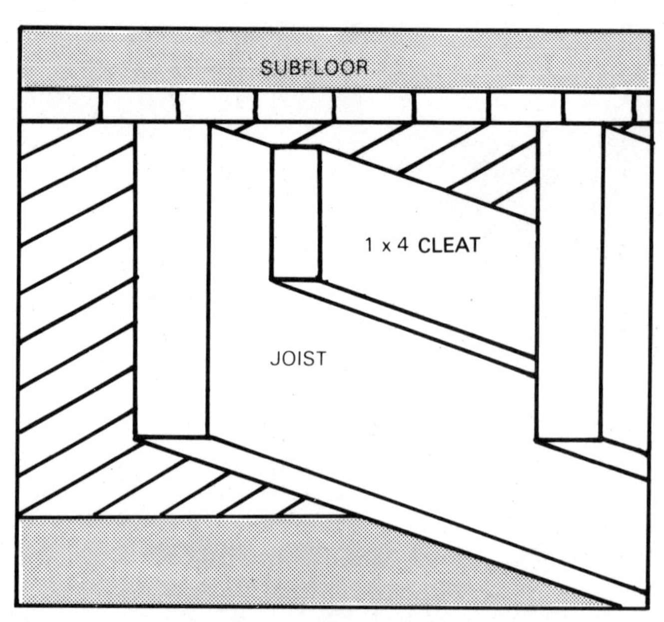

If several subfloor boards are loose, a 1 x 4-inch wooden cleat can be fastened to the joist.

against the loose boards when nailing.

3. If the subfloor is constructed of plywood sheathing, you may detect movement in the joints between panels. Use screws to attach a 1 × 4 to the two panels, centered over the joint. Movement will be eliminated.
4. Small squeaks between boards can often be eliminated by applying a lubricant to reduce friction. Graphite, floor oil, or mineral oil are suitable lubricants. Always use minimal amounts. If the lubricant proves inadequate, drive a few glazier points into the joints between squeaking boards.
5. If a joist seems loose or slightly warped, it can be strengthened significantly by nailing additional bridging between the problem joist and the two adjacent joists.

The importance of repairing the subfloor before applying a new floorcovering cannot be stressed enough. Hard materials such as ceramic tile, mosaic tile, quarry tile, wood parquet and rigid types of resilient tiles will crack or loosen when applied to an uneven or rough subfloor or underlayment surface. The first step in achieving a beautiful, well-installed floor is preparing the proper base.

The Finish Flooring Choices

The availability of a variety of floorcovering materials on the market today is staggering. Narrowing down the choices and selecting the most appropriate floorcovering may very well be the most difficult task when installing a new floor. The chart at the end of this chapter compares the various features of basic floorcoverings to help make your selection easier.

A floorcovering is a decorative element that should enhance the overall mood of a room. No single material is perfectly suited to the needs of every room in the home. The floorcovering should harmonize with furnishings, lighting, wallcoverings, window and door trim, ceiling, and overall color scheme. Color, texture, and pattern of the floorcovering play an important role in creating mood. Dark colors make a room seem smaller; lighter colors make it seem larger. Pattern design can accentuate furnishings.

Look at several floorcoverings at flooring showrooms. Ask the salespeople for specific information about durability, ease of installation, ease of maintenance, wearability, thermal insulation, sound conditioning, cost, type of underlayment required, materials necessary to install the floor yourself, and so forth. Try to anticipate your needs when you visit flooring dealers' showrooms.

Cost is usually an important factor; however quality floorcoverings generally cost only slightly more than those of lesser quality. If you consider the additional expense of replacement (when lesser quality materials wear out before quality materials would), the quality floorcoverings will prove more economical in the long run.

Flooring manufacturers are presently producing materials designed specifically for do-it-yourself, nonprofessional installation. They are easy to install with a few basic tools and require a minimum of care and maintenance.

Ceramic Tile

Ceramic tile includes glazed clay tile, quarry tile, mosaic tile, and unglazed tile. These practical flooring materials are suitable for any room in the house, but their most popular use is in the bathroom and kitchen. Although expensive and relatively difficult to install, they generally last for the entire life of the home.

The surface of ceramic tile is extremely hard and inflexible, so it reflects sound and requires a structurally sound subfloor. Its advantages outweigh its disadvantages considerably. Ceramic flooring materials are very durable, easy to clean and maintain, and, with new application techniques available from some flooring dealers, no longer require professional installation. Often ceramic tile can also be applied to walls and counters.

Ceramic tiles are made by pressing clay into shapes and firing them at high temperatures in kiln ovens. Tile is available in several styles and finishes and in a wide range of brilliant colors that will never fade, decorative patterns, and geometric shapes.

- Glazed tile—has a glaze that is applied to the clay tile prior to kiln firing. When fired, the glaze is actually baked onto the surface. Glaze colors are available in glossy (the most common), mat, satin, or dull finish. The glaze is somewhat susceptible to scratching. Glazed tiles range in size from 1 inch to 12 inches square.
- Mosaic tiles—are small glazed tiles, usually 1 or 2 inches square, mounted in groups on a backing sheet. This allows installation of large areas at one time. When set in place, grout is added between individual tiles.
- Quarry tile—is available either unglazed or glazed. Unglazed red quarry tile is most common. Quarry tile is easy to maintain, moisture resistant, and very durable. Common sizes are 4 × 8, 6 × 6, 8 × 8 or 12 × 12.
- Unglazed tile (or pavers)—relies on the clay for color; therefore, the color range is limited to earth tones. It is somewhat rough and semiporous. A sealant can be applied after installation for moisture resistance and ease of maintenance.

Resilient Flooring

Resilient flooring is a generic term for flooring that was traditionally known as linoleum. This type of flooring has experienced the greatest advancement and improvement in recent years. The development of synthetics and resins has created a large family of floorcoverings that is highly popular today. Resilient floorcoverings come in a virtually limitless variety of colors, patterns, textures, and designs.

Most resilient flooring is vinyl and relatively soft. Although this softness makes it more susceptible to dents and tears, it is less likely to crack when applied over a rough subfloor or existing tile floor. The softness also makes the material more pliable for easier installation. Cost varies considerably. Expensive materials tend to offer greater durability, wearability, and color retention.

If the existing worn floorcovering you plan to replace is not wood, ceramic tile, or carpeting, it is probably one of four types of resilient floorcoverings that were once in common use. Those four types are:

- Linoleum—was an excellent, popular floorcovering several years ago. It was the traditional resilient floorcovering used primarily in the kitchen. Linoleum is a blend of linseed oil, pigments, fillers, and resin binders bonded to asphalt-saturated felt. Linoleum provides fair wearability, easy cleanability, and excellent resilience. It is available in both tile and sheet form, but because it is no longer manufactured in this country, it is often as expensive as sheet vinyl flooring. It is no longer widely used nor easy to find.
- Asphalt tile—is a combination of asbestos fibers, limestone powder, and mineral pigments with an asphalt binder. It is relatively inexpensive. Dark colors are most common. Asphalt tile is brittle and less resilient than other floorcoverings. It is easily stained by grease, so use in kitchens is not recommended. It is best used on concrete slabs, making it a popular tile for basement recreation room floors. Asphalt tile has largely been replaced by the vinyls and is generally only available by special order in limited quantities, primarily for commercial use.
- Rubber tile—is made from synthetic rubber. It is long-wearing, moistureproof, resilient, and quiet. Rubber tile requires frequent waxing and buffing to maintain its lustrous, smooth surface. The surface is slippery when wet. Rubber tile was made primarily for industrial use and is no longer manufactured domestically.
- Cork tile—wears rapidly and is not impact-resistant. It is easily broken down by grease and alkali cleaners. Like rubber tile, it is no longer manufactured domestically.

Vinyl Flooring A vinyl floorcovering is basically plastic containing polyvinyl chloride (an oil derivative), resin binders, and pigments bonded to a felt or asbestos backing. Vinyls are easy to install and maintain. The designs and colors are bright, fadeless, and generally will not wear away. Vinyls offer outstanding resistance to wear, moisture, grease stains, and alkalies. They are softer and more flexible than previous resilient floorcoverings. Vinyl floorcoverings can be used in any area of the home but are most popularly used in the kitchen.

Vinyl floorcovering is available in sheet or tile form.

- Sheet flooring—is manufactured in large rolls up to 12 feet wide. The greatest advantage of sheet vinyls over tile vinyls is the absence of seams making more sophisticated styling possible. Sheet vinyls are generally more expensive and slightly more difficult to install.
- Tile flooring—is most commonly available in 12-inch squares; 9-inch squares are available in limited styles. There are two types of vinyl tiles, solid vinyl and vinyl composition. Vinyl composition tile has many of the same characteristics as sheet vinyl. Vinyl asbestos tile has asbestos

Ceramic tile blends nicely with the other design elements of this family/recreation room. The natural texture and look of the tile complements the wood paneling, plants and ceiling beams.

Resilient floor tiles create a durable, easy-to-clean surface for this dining and family room.

Wood flooring can create a wide variety of beautiful and elegant floor designs.

fiber fillers between two layers of vinyl. It is slightly less expensive than vinyl composition and is harder, less resilient, and more susceptible to abrasion and damage from household cleaners. Tiles have long been preferred by do-it-yourselfers for their ease of installation. In the past few years, tile patterns have become as beautiful as the designs of the more expensive sheet vinyls.

Vinyl floorcoverings, whether in sheet or tile form, are also available with other desirable features: they may be cushioned and have a no-wax finish and a self-stick backing. Installation of a vinyl floor does require some patience and care, but most homeowners have the necessary skills to install one.

Wood Flooring

Wood flooring has been the most desirable floorcovering installed in new homes for many years. Even with the advent of far less expensive synthetic materials that duplicate the rich tones and grain patterns of natural wood, wood flooring maintained its popularity. Wood has a natural beauty, warmth, and durability that is suitable in any room in the home.

Although both hardwoods and softwoods can be used for flooring, hardwoods are most commonly used because of their higher resistance to wear. The two most popular hardwoods are oak and maple. Others include birch, beech, and hickory. Oak has excellent grain pattern and character; maple is smooth, strong, and very durable. The open grain of oak generally requires more sealing than the even, uniform maple surface.

When properly sealed and finished with a durable finish coat, softwoods such as yellow pine, cedar, fir, or hemlock make excellent flooring materials. Softwoods are generally slightly less expensive than hardwoods.

All wood floors should be sealed and protected with a finish, usually varnish or one of the newer, more durable polyurethanes. These materials are resistant to stains, moisture, scuffs, and scratches. When a wood floor becomes worn looking, it can be refinished to look like new.

Wood flooring is expensive but, like ceramic tile, will often last longer than the home. Cost depends on type of wood, grade, saw pattern, and whether it is prefinished or not. Wood tends to shrink as it ages, particularly in hot, dry conditions; conversely, it can swell in humid conditions.

There are three basic types of wood flooring:

- Block flooring—is made of solid pieces of wood held together by some form of backing. Blocks are usually rectangular or square in shape and are available in several patterns, textures, and designs. The most common type of block flooring is parquet. Parquet has increased in popularity the past few years because it offers the natural beauty of a custom wood floor but is no harder to install than vinyl or ceramic tile. The most common size is the 6 or 12-inch square. Popular woods are oak, maple, teak, and cherry.

- Strip flooring—is the most common type of wood flooring. It is comprised of narrow wood strips of widths commonly between 1½ and 3 inches. The edges are tongue and grooved to provide tight uniform joints. Strip flooring is available finished or prefinished. Prefinished is more expensive but saves considerable time and effort. The most common woods used for strip flooring are oak, maple, and birch. Installation is difficult and may not be within the ability of the average homeowner; however strip flooring products for do-it-yourself installation are available.
- Plank flooring—is basically the same as strip flooring except that the boards are wider, ranging from 3 to 8 inches wide. In homes built in the eighteenth and nineteenth centuries, planks were attached to the subfloor with wooden pegs. Today pegs are simulated in the planks to give a colonial appearance. Floor planks may be hardwood or softwood; many planks have tongue-and-groove edges for easier installation. Installation of plank flooring is only slightly easier than strip flooring because fewer boards are needed to cover a given area.

Carpeting

Carpeting has traditionally been used primarily in the living or family room and bedrooms, but with some of the new hard-wearing, easy-to-clean materials available, it is being used even in the kitchen and bathroom. Carpeting is available in hundreds of fibers, styles, and colors with prices that vary considerably. The primary advantages of carpeting are appearance and comfort.

Carpeting can usually be installed over a clean, smooth existing floorcovering. Conventional carpeting has no backing and requires the installation of a pad beneath it to add comfort and prolong carpet life. Cushion-backed carpeting has a bonded rubber backing and requires no pad.

Carpeting can be made of one of several basic fibers. No one fiber is best, and combinations are available. Natural fibers, such as wool or cotton, have always been popular, but they are generally more expensive than the synthetics, such as polyester, nylon, acrylic, rayon, or polypropylene. The synthetics are usually less expensive, more durable, and eaiser to maintain. Choice of fiber should be governed by the expected traffic level in the area where it will be installed.

There are three basic types of carpeting: room-size rugs sold in standard sizes like 6×9, 9×12, and 12×15 feet; room-fit wall-to-wall carpeting sold on rolls, 6, 12, and 15 feet wide (roll carpeting usually requires professional installation); and carpet tiles in 12 and 18-inch squares. Carpet tiles are primarily used in commercial installations.

The textures produced by various manufacturing processes are called styles. Texture depends on the way the carpet is made—whether by tufting, weaving, knitting, needle punching, or flocking. There are four basic styles:

- Sculptured-pile carpeting—has a rich, soft traditional texture with several surface levels created by using varying lengths of pile. Common fibers are wool, nylon, and acrylic.
- Shag carpeting—is generally less expensive than sculptured pile because the fibers are not as dense, and it is also not as durable. Nylon is preferred because of its easy cleanability. Look for tightly twisted full strands of fiber of varying lengths.
- Plush carpeting—is very luxurious looking because the pile is so closely woven. Acrylic fiber gives the carpeting a very soft appearance. Its shading can create the illusion of a rich two-color design. Plush carpeting is expensive.
- Level-loop carpeting—has tightly constructed loops of uniform length fiber. This carpeting is easy to maintain and clean; it is one of the most durable as well.

Carpeting offers warmth and comfort to this below-grade recreation area.

Comparing Types of Finish Flooring

TYPE	RELATIVE COST	EASE OF INSTALLATION	EASE OF MAINTENANCE	GENERAL USES	DURABILITY	RESILIENCE	SAFETY
WOOD							
Wood Parquet or Block	High to medium	Moderately easy	Good, dust mop and wax	All areas except bath and utility areas	Excellent, should last for life of the house	Poor	Good, can be slippery when waxed
Wood Strip or Plank	High	Difficult	Good, dust mop and wax	All areas except bath and utility areas	Superior, should last the life of the house	Poor	Good, can be slippery when waxed
CARPET							
Shag Carpet	High to medium	Moderately difficult	Good, vacuum weekly	All dry areas of the home	Good, can fade or mat with time	Good to excellent	Excellent, insulates and softens surface
Sculptured or Loop Carpet	High to medium	Moderately difficult	Good, vacuum weekly	All dry areas of the home	Good, tends to wear in high traffic areas	Excellent	Excellent, insulates and softens surface
Carpet Tiles	Medium to low	Moderately easy	Good, vacuum weekly	Primarily commercial	Fair, tend to curl at edges	Good	Very good
RESILIENT							
Sheet Vinyl	High to low	Moderately easy	Very good, sponge mop, rinse, wax	All areas of the home	Very good, can fade with time	Excellent to fair	Good, can be slippery when wet
No-Wax Sheet	High to medium	Moderately easy	Superior, never needs waxing	All areas of the home	Very good, can fade with time	Fair	Good, can be slippery when wet
Cushioned Sheet Vinyl	High to medium	Moderately easy	Excellent, stain resistant	All areas of the home	Very good, can fade with time	Very good	Good, can be slippery when wet
Vinyl Asbestos Tile	Low	Easy	Very good, sponge mop, rinse, wash	All areas of the home	Good to superior	Fair	Good, can be slippery when wet
Vinyl Composition Tile	Medium to low	Easy	Excellent, requires little maintenance	All areas of the home	Superior	Fair	Good, can be slippery when wet
CERAMIC TILE							
Ceramic Tile	High to medium	Moderately difficult	Superior, occasionally damp mop and clean	Bathrooms, kitchens	Superior, should last for life of house	Poor	Good, can be slippery when wet
Mosaic Tile	High to medium	Moderately difficult	Superior, occasionally damp mop	Bathrooms, kitchens, halls, utility rooms	Superior, should last for life of house	Poor	Good, can be slippery when wet
Quarry Tile	High to medium	Moderately difficult	Superior, occasionally damp mop	Family rooms, kitchens, bathrooms, foyers	Superior, should last for life of the house	Poor	Good, can be slippery when wet

Measuring Area and Sketching a Floor Diagram

Before selecting or ordering any floorcovering material, determine the exact area that you plan to cover. Drawing a scale map of the area to be covered will enable you and the floorcovering dealer to know the precise amount of floorcovering and trim material required. Note all measurements on the map as accurately as possible. Indicate all physical facts of the floor area as well. Buying too little or too much of a floorcovering is an obvious waste of money and materials that should be avoided.

Be slightly generous in the amount of floorcovering you order. You need enough materials to cover the entire area plus a little more to allow for miscalculation, damage during installation and replacements should materials break or become damaged. Usually about a 5 percent additional amount is sufficient. Since most floorcovering materials are relatively expensive, why pay for more than you will use?

Drawing the Map

To make an accurate scale drawing, you will need the following: graph paper ruled in quarter-inch squares and large enough to accommodate a sketch of the entire area to be floored, pencil, tape measure, a self-chalking chalk line, and several pieces of chalk. If you do not have help making the necessary measurements, the end of the tape can be secured with a tack or small nail to prevent possible shifting of the tape.

The recommended procedure for drawing a floor diagram is:

1. Measure the longest dimension of the area first.
2. After taking the measurement, draw a line on the graph paper representing the line (A to B). For large areas, one square on the graph paper can equal 1 foot. For smaller areas, one square

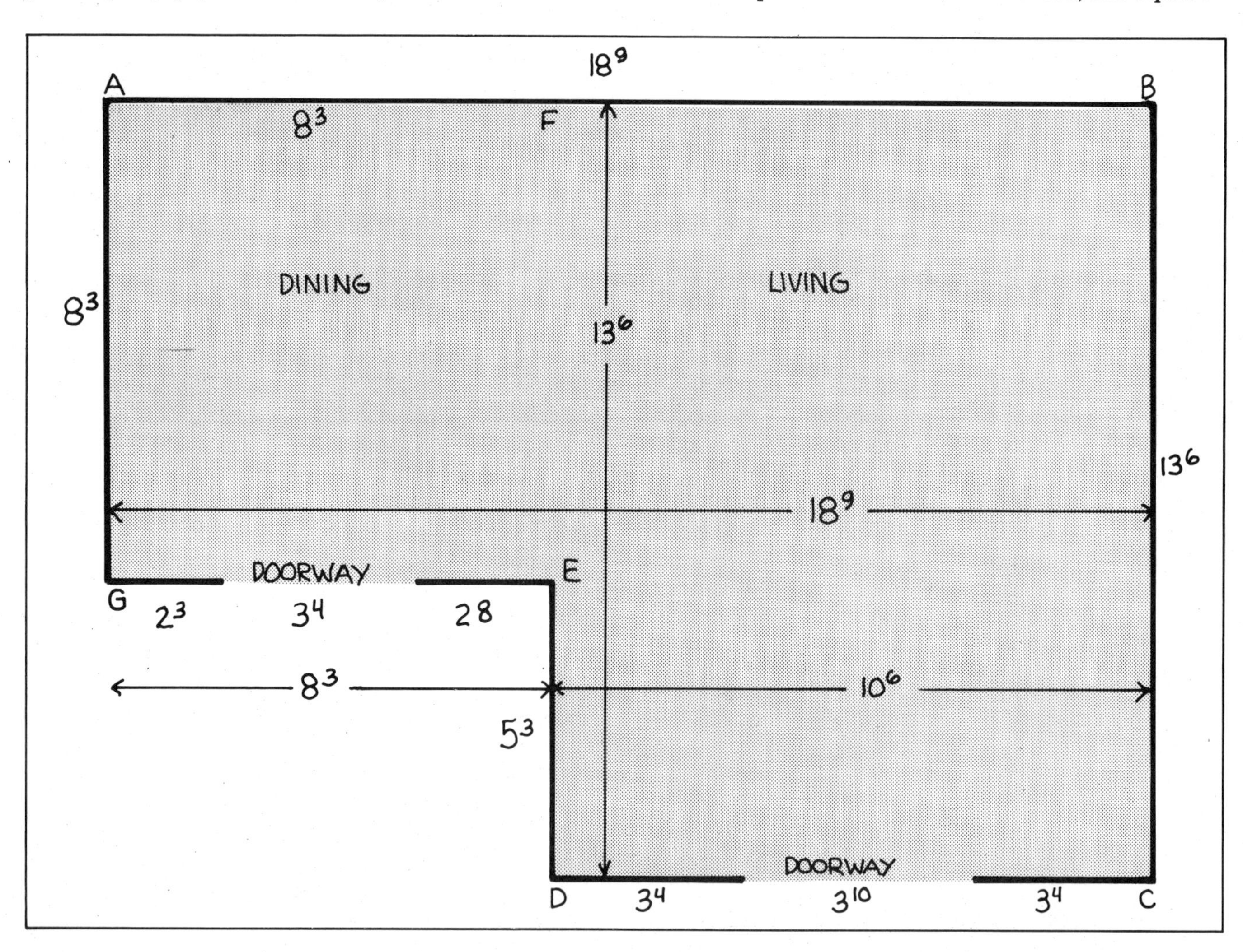

can equal 6 inches. Center the longest dimension near the top of the graph paper to be sure the entire area will fit on the paper.

3. Measure the longest adjacent wall at a right angle to the starting wall. Draw it on the graph paper using the same scale (B to C). Indicate the exact measurement of the wall outside of the line on the graph paper for clarity.
4. Measure the adjacent walls and draw them in proportion on the graph paper. Indicate all exact measurements outside lines.
5. Be sure to measure all short walls, doors, passageways, closets or other features. First measure the entire wall length (C to D) and indicate this line on the graph paper. Now indicate the broken measurements—the two short walls from corner C to doorway and corner D to doorway, and the width of the actual doorway. The total sum of all broken measurements should equal the measurement of the total wall length. If they are not equal, measure again. Continue measuring all remaining walls in this manner.
6. It is not uncommon for rooms to have one or more walls that are angled. This is easily discovered by measuring the lengths of parallel walls. If lengths of parallel walls are identical, the room is square. If they are not, the room is angled. Although wall DE is only 4 feet long, take an overall measurement from corner D to opposite wall AB. If the measurement from D to the junction with AB is longer or shorter than the measurement of opposite parallel wall BC, then wall CD is angled. Since D to AB is 6 inches longer than B to C, the floorcovering order must be based on wall D to AB. If it was based on B to C, there would not be enough floorcovering material to cover the actual area.
7. Measure the next wall (E to F). Again indicate all broken measurements and be certain the sum equals total wall length. Measure the length of parallel walls as well. Measure from corner F to the junction with the opposite wall (B to C). If this measurement equals the measurement of wall A to B, the room is square. If not, the room is out of square. Always take time to verify squareness by comparing overall measurements of parallel walls.
8. The final wall is F to A. Measure and indicate the wall on the drawing.
9. Measure and draw the distance the floorcovering will go into doorways and closets. Failure to do so will result in unnecessary seams. Every detail is important.

When all details have been clearly and accurately indicated on the scale drawing, take it with you when you shop. It will give the dealer all the necessary information needed to estimate the amounts of materials required for the project. Planning is important to any home improvement project and begins with accurate measurements.

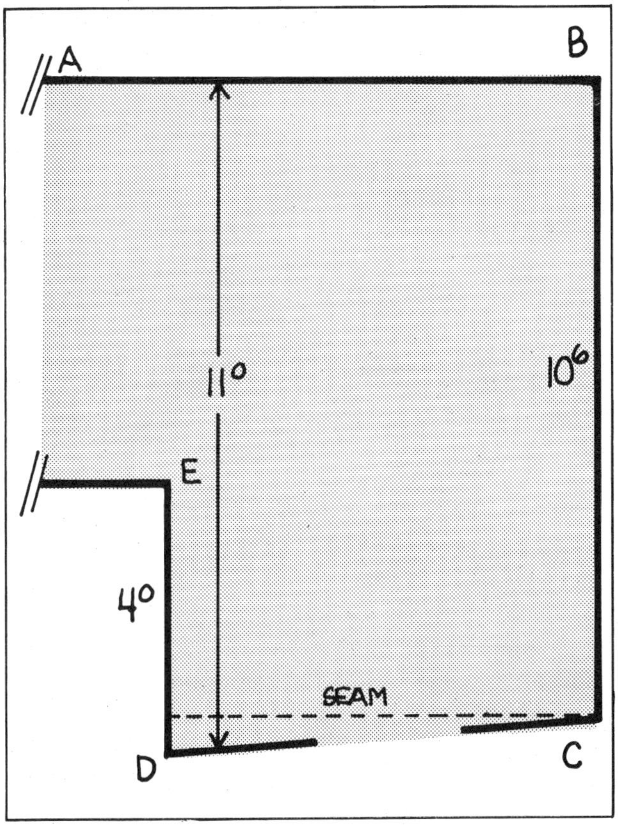

These walls are not parallel. If the floorcovering order was based on wall BC rather than D(AB), a seam would occur near the doorway by wall CD.

Ceramic Tile

Ceramic tile is one of the oldest, most popular floorcovering materials available. It has withstood the test of time with amazing success. Beautiful ceramic tile floors and walls can be found today in cathedrals, temples, and palaces built hundreds of years ago. Early Greek, Roman, and Egyptian cultures used ceramic tiles in the construction of important city buildings.

Very few floorcovering materials offer both the decorative and functional qualities of this durable material. That is why ceramic tile deserves careful consideration whether you are building a new home or remodeling an existing one. When properly installed, ceramic tile will last the entire life of the home and will actually add to its value at the time of sale. Ceramic tile is impervious to heat, soil, moisture, and mars. It is easier to maintain than most other floorcoverings.

Ceramic tile is an earthborn material that is toughened in the 2,000-degree heat of a kiln. It will stand up to wear longer than most other natural or synthetic materials. It will not burn, tear, cut, dent, warp, or blister.

In older homes, ceramic tile was primarily restricted to use in the bathroom because of simple economics. Ceramic tile is relatively expensive floorcovering material, and in the past professional installation was generally required. Materials and labor made the cost of ceramic tile too prohibitive to be used throughout the home. Today, however, installing a ceramic tile floor can be a highly rewarding, moderately easy do-it-yourself project because manufacturers have tailored their products for easier installation. Easy-to-install tiles, specific instructions, self-spacing tile sheets, and improved adhesives and grouts have placed installing a professional-looking ceramic tile floor within the scope of the careful, patient do-it-yourselfer.

A Wide Variety of Choices

Selecting a ceramic tile that best meets your needs and tastes may be the most difficult step in installing a tile floor. Ceramic tile is available in literally thousands of colors, patterns, sizes, shapes, and surface textures. It is an incredibly versatile material that can be adapted to virtually any mood, appearance, or situation. By selecting several colors or styles, the homeowner can arrange the individual tiles into a stunning customized pattern or mosaic. The visual impact of a beautiful, unique design can be an expression of creativity and personality.

The colors of ceramic tile are permanent and will not fade. Colors range from bright, brilliant hues that can give a room depth and glow to rich natural earth tones that can give a room added warmth and comfort. Tiles are available in plain solid colors or highly decorative patterns. Color depends on how the tile was finished. Every tile regardless of color, shape, size, or texture is finished in one of two ways—glazed or unglazed.

All ceramic tile, whether glazed or unglazed, is essentially a slab of hard-baked clay that has been fired twice in high-temperature kilns. Glazed tiles have a hard, colorful surface applied to the clay body. After the initial firing, a glaze can be applied to the cooled surface of the clay body known as the bisque. It is this glaze when fired the final time that gives glazed tile its highly colorful surface. Glazes are available in several finishes: dull, mat, satin, or high gloss. Glaze makes the surface of ceramic tile moistureproof.

Unglazed tiles derive their color from the clay itself. Their colors are somewhat dull and are consistent throughout the tile. Clay is available in several natural colors or can be made more colorful by adding pigment prior to molding and firing. The dull, semiporous surface of unglazed tiles can be made more glossy, and subsequently more moistureproof, by applying a wax, sealer, or finish to the tiles or tiled floor surface. Unglazed tile offers greater traction when wet.

Ceramic tiles come in many shapes, with squares and rectangles being the most popular. Several geometric shapes are available including hexagonal, octagonal, and a few other unique styles. Tiles are also available in many thicknesses and sizes. Floor ceramic tile is thicker than wall tile, usually ranging from 3/8 to 3/4 inch thick. The added thickness is for increased strength and durability. Never use wall tile on floors. Floor tile sizes are generally larger than wall tiles, with the most popular sizes being 1×1, 4×4, 6×6, 8×8, and 12×12 inches. The tiles can be purchased individually or by the sheet, particularly the smaller tiles. Sheets of floor tile range in size from 6×12 to 12×24 inches. Most sheet tile is self-spaced for easy installation.

There are three basic types of ceramic tile: quarry tile, ceramic tile, and mosaic tile. All three types are similar in composition and manufacture but differ in size, shape, and the manner in which they are installed. Each type will be discussed.

Quarry Tile Quarry tile is made by the extrusion process from natural clay or shale. These tiles are

The quarry tile floor in this greenhouse offers natural beauty and an easy-to-clean surface.

generally the largest, most durable ceramic tiles. They are available glazed or unglazed, although unglazed, in the natural earthy tones of clay ranging from tan to dark brown, are most commonly used. Because quarry tiles are so durable they make an excellent floor in high-traffic areas and can be used outdoors as well as indoors. They are water-resistant, so they can be used in the bathroom, kitchen, or patio area. The slightly rough surface offers excellent traction when wet. Glazed quarry tiles have many inherent shade variations and a long-wearing tough surface that resists scratches and abrasion. They are also cut easily with standard cutting tools. European designs and colors are also available. Unglazed tiles, usually an earthy red color, are frostproof, so they can be used outdoors in cold climates. The tiles range in thickness from 3/8 to 7/8 inch. Common surface sizes are 4×8, 6×6, 8×8 and 12×12 inches. The rectangular-shaped tiles can be installed in a wide variety of patterns and designs. Other shapes include Moorish, hexagon, octagon, and ogee.

Ceramic Tile Ceramic tile is an excellent floorcovering that is durable, colorfast, and very easy to maintain. Ceramic tile is glazed and comes in a seemingly endless variety of decorative patterns and colors that will remain bright and shiny for the life of the tile. The range of colors and patterns affords great design flexibility. Many sizes and shapes are available in thicknesses ranging from 3/8 to 5/8 inch. The most common sizes are 3 × 3, 4 × 4, 6 × 6, and 8 × 8 inches. The tiles are available in two finishes, high gloss for unmatched beauty and mat or textured for better slip-resistance when wet. Ceramic tile is available as single tiles or, from some manufacturers, as sheets of several tiles.

Mosaic Tile Mosaic tile has a backing sheet of thread mesh, plastic, paper, or silicone rubber that allows several individual tiles to be attached together in predetermined spacing. This makes it easy to install and lends itself to round and other nonflat surfaces. It is a very versatile type of ceramic tile. Ceramic mosaics are every bit as colorful, tough, and durable as the two other types. Finishes range from natural and textured for good traction to mat or gloss glazes for bright decorative colors. Certain mosaic tiles are baked at extremely hot temperatures to create porcelain tiles, harder and denser than conventional ceramic tile. Some mosaics are suitable for outdoor installation. Mosaic tiles are usually 1/4 to 1/2 inch thick and range in size from 1 x 1 to 3 x 3 inches. Sheet sizes are approximately 1 foot square, although larger sheet sizes may be available from some manufacturers. Common mosaic patterns are squares, rectangles, and hexagons.

Where to Use Ceramic Tile

In recent years ceramic tile has experienced an amazing increase in popularity. Once used primarily in bathrooms or entryways, ceramic tile today is used throughout the home, indoors and out. Because of its unsurpassed durability, ceramic tile is an excellent floorcovering material, particularly for high-traffic areas. Since ceramic tile is also a strong decorative element, it is suitable for use on walls in many rooms. Ceramic tile is easy to clean and is waterproof, making it perfect for use on kitchen countertops or bathroom tub enclosures. Because tile is fireproof, it can be used on a fireplace facing or hearth. A patio of frostproof ceramic tile can easily withstand the elements to make an outdoor recreation area even more attractive.

Although many room designs featured ceramic tile because of its durability, it is now being selected because of its versatility and elegance. Always an excellent investment because of easy care and permanence, ceramic tile offers more design possibilities today than ever before. Practically any design idea or decorating plan can be satisfied with the wealth of colors, moods, patterns, and textures of ceramic tile.

Of all the living spaces in the home, the bathroom is one of the most used and probably the most permanently arranged. Until recent years bathroom decor was a matter of compactness, efficiency, and function, not beauty. Today, however, bathroom redecorating and remodeling is one of the most popular projects for the do-it-yourselfer.

Even the smallest of bathrooms can be redecorated to appear larger, more relaxing, and as colorful and enjoyable as any other room in the house. The bathroom floor is a natural place for ceramic tile's durability, cleanliness, and beauty. The design you select for the floor can be used on walls, countertops, and tub enclosures as well. The look of today's bath is focused on creativity, color, and endless new possibilities in pattern and texture.

The entryway has long been a popular area for ceramic tile. The floorcovering of an entryway must be durable since it is probably one of the highest traffic areas in the home. It must also be easy to clean since this area tends to become dirty faster than most others. Entryway floorcoverings must be functional, but they also have another very important function—they must welcome visitors into the home and enhance the overall decorative scheme of the home. Depending on the interior decoration of your home, the entryway can be a simple tile surface or a creative mosaic graphic design. Both designs are functional, but they also

Quarry tile can be used outdoors or in greenhouses.

This vacation home is floored entirely with ceramic tile for a clean, beautiful appearance.

Ceramic tile is perfect in the kitchen where an easy-to-maintain surface is essential.

Bathrooms, too, benefit from the easy-maintenance feature of ceramic tile.

Ceramic tile is not limited to use on floors. Countertops, tub enclosures, walls and even ceilings can be covered with ceramic tile, particularly in areas where moisture can cause problems.

complement the surrounding decor of their respective homes.

The kitchen is quickly becoming one of the most popular areas of the home for ceramic tile. It has a brilliant clean face that can brighten any kitchen, and the natural look and feel of tile enhances the rich, warm tones and textures of wood cabinets, paneling, and dining area furniture. Ease of maintenance is an important consideration when selecting a kitchen floorcovering, and ceramic tile surpasses most other floorcoverings in this important regard. Ceramic tile is also perfectly suited for use on kitchen countertops and backsplashes.

Ceramic tile is an excellent floorcovering material for several other areas in the home. Its durability and wearability qualities make it perfect for use in sunrooms, attached greenhouses, game rooms, basement recreation rooms, or workrooms. Tile helps create an outdoor look indoors. Wood paneling is very compatible with ceramic tile.

Frostproof ceramic tile's attractive permanence makes it ideal for outdoor settings. Whether for patios, walks, steps, or accents, it is an excellent choice. Although this is an interior flooring book, these examples stress the great versatility and durability of ceramic tile.

Preparing the Subfloor

A proper subfloor for ceramic tile is crucial. If you install a ceramic, mosaic, or quarry tile floor over anything but a clean, level, dry, and structurally sound surface, you are inviting trouble. A successful ceramic tile floor requires careful subfloor preparation. Without proper preparation, any ceramic tile floor, no matter how well installed, will soon begin to deteriorate.

The entire working surface area must be prepared. First, remove furniture, furnishings, rugs, and any other objects from the area to expose as much of the floor surface as possible. Remove the baseboard trim from around the room, if possible, without damaging the adjacent wall. Number each piece so it can be placed into its original position after the new ceramic tile floor is completed. If the baseboard trim is impossible to remove, then remove the shoe molding and tile to the base of the baseboard trim. When removing any floor-to-wall trim from painted walls, first score the painted wall by running a sharp knife along the top of the baseboard trim to prevent the paint from chipping. Also remove any other obstructions such as heating-cooling registers and floor electrical switch plates.

In the bathroom, floor-mounted toilet fixtures must be removed. This is a relatively easy task. Shut off the water-supply valve, flush the toilet, remove any excess water from the bowl and tank with a sponge, and disconnect the water-supply line. Remove the nuts from the bolts holding the toilet base to the drainpipe. Remove the fixture and set on newspapers in a nearby room.

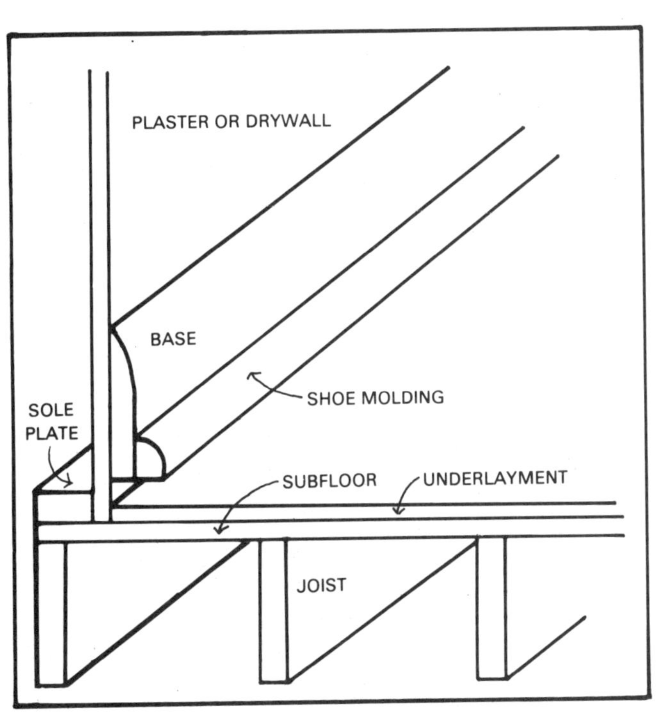

The doors which open into the area must also be removed. Since the tile and mortar may raise a floor by at least ¾-inch, the door may have to be trimmed to clear the new floor height. Lay two loose tiles, one on top of the other, on the floor next to the closed door. Draw a line along the door across the top of the tiles. Unhinge and remove the door. Cut the bottom of the door off at the line.

With as much of the floor surface exposed as possible, examine the condition of the existing floor or subfloor surface. First examine the subfloor from below as described in the chapter on basic floor construction. Make any necessary repairs to the structural framework.

A ceramic tile floor can be installed on several types of new or existing subfloor backings, or it can be installed over a variety of existing floorcovering materials. Again, the key is a level, dry, strong, sound surface. The surface must not have any noticeable give to it, or the tile may loosen or crack under stress.

The Existing Subfloor

Ceramic tile can be applied over most conventional residential subfloors, provided the backing material is sound and strong. If the present backing is springy or uneven, a new backing must be applied before installing ceramic tile. The most common subfloors are plywood panels, wood boards, or concrete slab.

Plywood Panels Most homes today have subfloors constructed of ½ or ¾-inch plywood panels with staggered joints. Particleboard is often substituted for plywood. Be certain all panels are securely fastened to the joists, with no protruding nailheads. If you detect any loose areas, nail them to the joists with ring-shank nails. Remove all raised surface imperfections with a scraper or putty knife. Nailheads, gouges, dents, and cracks will be filled with the mortar that will hold the ceramic tile in place. The mortar will also cover any slightly uneven areas. If the existing subfloor must be replaced, use ¾-inch plywood. Panels should be nailed at right angles to the joists.

Some plywood subfloors may not be rigid enough for ceramic tile. If the individual panels give when you walk on them, the subfloor must be reinforced with another layer of plywood. Use ⅜, ½, or ¾-inch sheets to create a total subfloor thickness of 1½ inches. Most panels are available in a 4 × 8-foot size. Select an exterior or underlayment grade plywood. Stagger the new panels perpendicular to the original subfloor. Lay the new panels at right angles to the old. Leave ⅛-inch space between panels. Fasten with ring-shank nails driven through the original subfloor and into the joists.

Organic adhesives and epoxies work best on plywood surfaces. Some adhesive directions recommend that the wood be primed or sealed to increase water and moisture-resistance and to strengthen the bond between adhesive and subfloor surfaces.

Wood Boards In many older homes, the subfloor is made of individual boards that are 4, 6, or 8 inches wide, nailed at a 45-degree angle to the joists. Nail down any loose planks. Because ceramic tile is very heavy, a plank subfloor is generally not strong enough to support the weight. Warping and the subsequent cracking of tile can occur. Before installing tile, cover the plank subfloor with plywood or another form of underlayment backing as described in the paragraphs above.

Concrete Slab Concrete is probably the best base for ceramic tile, both indoors and out. The concrete must be fully cured, clean, and absolutely dry. It must also be structurally sound and reinforced with metal rods or mesh. Any cracking of the concrete will cause the ceramic tile floor to crack. If the present concrete surface is unsatisfactory, a new layer of concrete can be poured over the existing surface. Be sure to follow the adhesive manufacturer's recommendations for concrete subfloor preparation. Concrete must cure a minimum of 28 days before ceramic tile is laid over it. The concrete must also be totally free from moisture. Check this by covering an area with plastic, sealing the edges with duct tape. If there is any condensation on the underside of the plastic after several days, the concrete needs additional time to dry. After you are certain the concrete is thoroughly dry, sweep the slab to remove dirt and dust. Do not wet mop the floor. If there are spots that have waxy or greasy film, clean the area with a chemical cleaner. Patch any uneven areas and chip away any rough irregular areas. Apply asphalt primer to the surface and allow to thoroughly dry. If the adhesive is cement-based mortar, all irregularities must be filled with concrete patching compound. Apply mastic underlayment if you plan to use mastic. More extensive information regarding concrete subfloors can be found in the wood floorcoverings chapter.

Existing Floorcovering

An exposed floor is not always required for ceramic tile installation. Ceramic tile can be installed directly over several existing floorcoverings if the surfaces are structurally sound, level, and clean. The existing surface cannot be cracked, broken, irregular, or loose. If the floorcovering is in poor condition, however, it must be removed. The existing floorcovering should be removed also if it is imperative that the new ceramic tile floor be the same height as adjacent floors.

Hardwood Floors Strip, plank, and parquet floors are suitable bases for ceramic tile provided the floor is level and sound. The finish should be removed with a floor sander to achieve a better bond between wood and adhesive, particularly if you plan to use epoxy.

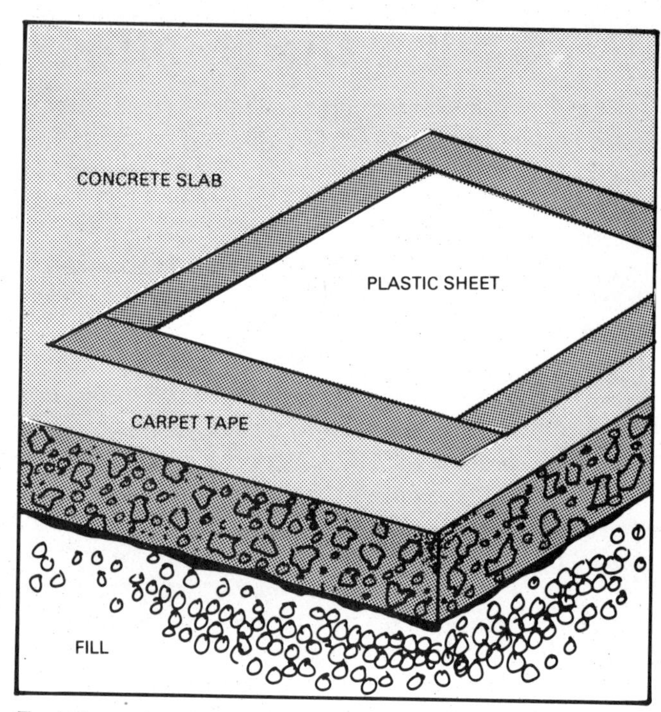

Test the moisture content of a concrete slab by fastening a sheet of plastic to the slab with duct or carpet tape.

Some organic adhesives may be compatible with certain wood-floor finishes. Check the instructions on the adhesive container. Some tile manufacturers recommend covering hardwood floors with an underlayment of plywood because the moisture in the adhesive may cause warping. Hardwood strip flooring is not as structurally sound as plywood panels either. Consult your tile dealer for accurate information concerning hardwood flooring as a subfloor for ceramic tile.

Resilient Floors Either cushioned sheet or tile resilient flooring is too soft and springy to make a suitable base for ceramic tile. Resilient flooring must be rigid, in good repair, and attached securely to the subfloor. If the resilient flooring is not in good condition, it must be removed from the subfloor; or, if additional floor height is acceptable, plywood panels can be nailed directly over the existing floor to create a new subfloor surface. If it is possible to apply ceramic tile over an existing resilient floor, use an organic or epoxy-base adhesive for best results. Your tile supplier can recommend the best adhesive for a specific resilient material.

Ceramic Tile Floors If an existing ceramic tile floor is clean, level and adheres well to the subfloor, new ceramic tile can be applied over it. Remove any loose, cracked, or broken tiles and replace with similar tile. Always look for water damage—broken or missing grout, loose tiles, and damaged subfloor or underlayment. Clean the tile thoroughly by lightly sanding the surface to remove wax, dirt, mildew, or mineral buildup. The sanding action will also scuff the surface to provide better adhesion. After sanding wash with clean water.

Installing Ceramic Tile

Step-by-step installation instructions and photographs for the three basic types of ceramic tile are presented in this section. The instructions for installing ceramic tile are detailed first. Since several of the installation steps for quarry and mosaic tile are similar to those for ceramic tile, read first the installation information for ceramic tile. If you plan to install quarry or mosaic tile, refer to the instructions for the specific tile for more detailed information.

Before proceeding make sure the subfloor is properly prepared and that all area measurements are accurate. Take the scale drawing of the area with you to the tile dealer. Purchase enough tile and trim to complete the job. The tile dealer can help you determine the exact amount needed. Remember, purchase 3 to 5 percent more tile and trim for insurance against mistakes and for future use if tiles become cracked, stained, or broken. Check each carton of tile and trim to make sure color and /or pattern are uniform.

Several specialized tools and materials are required to install ceramic tile. Most tile dealers will sell or loan the more specialized tools. Other installation materials, such as adhesive and grout, should be purchased from the same dealer that sold you the tile. This should ensure compatability between materials.

Adhesives

The traditional adhesive for ceramic tile is the cement mortar bed. Individual tiles have to be soaked before being laid in a thick bed of mortar. Professional contractors have used this method for many years; but for the average do-it-yourselfer, laying a good mortar bed can be difficult. With more people doing their own ceramic tile installation, many manufacturers have developed adhesives designed to simplify installation and save time. These new adhesives are called "thin-sets" because they are only 1/8 inch thick compared to a 1 to 1½-inch mortar bed.

Depending on type of tile and subfloor surface, one of three thin-set adhesives should be used. Consult your tile dealer for the proper adhesive for your particular tile and subfloor. The three basic types are:

Mastics Mastics are organic, pastelike adhesives that can be applied over existing tile, masonry, concrete, or plywood subfloors. One type of mastic is highly water resistant, so it is often used in bathrooms, basements, and other damp areas. Some of these mastics are formulated with a flammable solvent that can irritate skin and the respiratory system. Work only in well-ventilated areas and wear gloves. Another type of mastic is less dangerous to use, but it should be used in areas that are relatively free of moisture for best results. These mastics have a latex base, making cleanup much easier. Purchase the mastic best suited for the intended use.

Epoxy Adhesives Although these newer adhesives are generally expensive, epoxy-base adhesives offer exceptional bonding strength on almost all subfloor surfaces, even an existing resilient or hardwood floor. Epoxies are mixed from two or three materials just prior to application. Drying time depends on room temperature. Because epoxies offer a strong bond, cleaning excess adhesive from set tiles can be tedious. There are two basic types of epoxy-base adhesives: regular epoxy adhesives and epoxy mortars. Epoxy mortars have the greatest bonding strength and should be used where greater resistance to chemicals is essential. Wear rubber gloves to prevent skin irritation. Epoxy mortars are somewhat more difficult to use.

Mortars Mortars, either dry-set or latex-Portland cement, are cement-based adhesives that are best used on a concrete, ceramic, or masonry surface. They are not suitable for use on a wood subfloor because it may swell even if sealed and primed. Cement adhesives do not adhere well to synthetic materials, such as vinyl and linoleum. They are not flammable, do not irritate the skin, and can be cleaned easily with water. Both types of mortars must be mixed with sand. Some manufacturers offer premixed varieties that have to be mixed only with water. Dry-set mortars are generally mixed with water according to package directions. Mix, allow to stand 15 minutes, and then remix before application. Tiles do not have to be soaked prior to this type of installation as they do for the traditional mortar bed. Latex-Portland cement mortars are mixed with liquid latex prior to application, making them more water-resistant than dry-set mortars. Again, follow the instructions on the adhesive's package.

Grouts

The other material that must be purchased when you order tile and adhesive is grout. Whereas adhesive secures ceramic tile to the subfloor backing, grout fills the joints between tiles to keep out dirt, moisture, and other matter than can damage the adhesive bond and backing. Grout also gives definition to the ceramic tile pattern by providing contrast to the tile colors. Grout must be applied correctly because even tiny cracks will permit moisture to penetrate.

Grout is available in several colors, or color can be added with special pigments. Select a color that best matches that of the tile or the dominant color of the tile. Avoid light-colored grouts in areas where the tile surface may be subjected to staining. There are three basic types of grout: silicone grout, epoxy grout, and cement grout. Base your selection on the following factors:

- the type of tile you are installing. Many manufacturers recommend a specific grout for their products.

- the type of adhesive used to bond tiles to the subfloor.
- conditions of the area where the tile and grout will be installed (damp, dry, subject to chemicals and stains, and so forth).
- the width of the spacing between tiles.

Silicone Grouts Silicone rubber grout most closely resembles caulk. It is a flexible material that withstands drastic temperature changes, resists moisture or water, and will not mold or mildew. It is also applied in the same manner as caulk. Use silicone rubber grout where surfaces tend to shift or where expansion joints are required.

Epoxy Grouts Epoxy grouts are expensive and are very difficult to remove from the tile surface once they begin to harden. But the extra expense and time required for an epoxy grout is well worth it if you intend to use it in a high-traffic area or where chemical or water resistance is necessary. Epoxy grouts have incredibly strong bonding properties. They are available in several tones ranging from white to dark brown.

Cement Grouts These grouts are the most often used type of grouts because they offer hardness, uniformity of color, and flexibility. Cement grouts have a Portland cement base. The key to a successful cement grout application is selecting fresh material. The grout should be powderlike and lump-free. You will need to choose from four types of cement grout. Dry-set cement grouts are similar to dry-set mortar. The grout is simply mixed with water and applied over the tile. Latex-Portland cement grout is similar to latex-Portland cement mortar in that a cement-sand mixture is mixed with liquid latex. Sand-Portland cement grout is a mixture of sand, cement, and water, with the ratio of materials determined by the width of the spaces between tiles. Wider joints require a stronger grout. The more sand, the stronger the grout. Commercial Portland cement grout is yet another type. This type is the most difficult for the do-it-yourselfer to apply because the tiles must be wet when the grout is applied, and the grout requires a special damp-curing method.

Tools

Several tools are required for the proper installation of ceramic tile. Fortunately, most of these tools are common household general-purpose tools; however, some specialized tools are required. Since it is not economical nor practical to purchase these tools, check with your tile dealer to see if they can be loaned or rented. If not, check with local rental centers.

Assemble all tools prior to beginning the installation. The tools you will need to lay ceramic tile are:

- General purpose tools—claw hammer, pliers, nail set, putty knife, electric drill, and caulking gun
- Measuring tools—straightedge, level, carpenter's steel square, steel measuring tape or folding rule, contour gauge, and a chalk line and chalk
- Tile-cutting tools—tile nippers or glass cutter or commercial tile cutter, rod saw blades, and masonry drill bits
- Specialized tools for ceramic tile installation—notched trowel for adhesive application, rubber-faced float for grout application, jointer for smoothing grout joints, and a rubbing stone to smooth edges of cut tile
- Other miscellaneous tools and supplies—buckets, sponges, rubber gloves, cleaning pad, and grout sealer

Assemble all materials, supplies, and tools. Make sure you have everything necessary to complete the project. Clean tile if it is dusty or dirty. Dust can weaken the bond between adhesive and tile. Read carefully all directions and review the sequence of the installation procedure before actually laying the tiles.

Mark Working Lines

The importance of careful planning was stressed in the chapter on estimating materials and making a scale drawing. Establishing accurate working lines on the surface to be filled will save time and effort and ensure a professional-looking ceramic tile floor.

Ceramic tile can be installed two basic ways, working from the center of the area or starting at one wall. Working from the center of the area is the preferred method if the room is not perfectly square (as is the case with many rooms) or if you desire a decorative or symmetrical design that should be visually centered in the room. This method is best suited to the do-it-yourselfer. It can be used with any type of flooring, including resilient tile or wood parquet squares, that is set in a thin-set adhesive. Generally, this method will result in cut tiles along all walls. The other method, starting at one wall, usually results in only two walls with cut tiles; however, this method works only if at least two walls are perfectly square and if none of the walls are irregular. It is a slightly more complicated procedure and is better suited to a mortar bed.

Establishing and using accurate working lines is not difficult. The instructions that follow explain how to mark working lines for the installation of ceramic tile starting at the center of the project area and working toward the walls. If the room is not square or rectangular, position the working lines in the major area of the room.

1. Working lines are not lines paralleled to the walls of a room. Few rooms have perfectly square corners, so a room that appears to have parallel walls actually may not. Measure the width of each wall around the area to be tiled. Indicate the center point of each wall on the floor next to the wall.
2. Stretch a chalk line between the center points of opposite walls. Stretch tightly, lift the line, and let it snap.
3. Follow the same procedure for the other walls. Before snapping the second line, place a carpenter's square at the intersection of the first line and the chalk line. The intersection of the first line and chalk line must be a 90-degree angle. If the two lines will not cross at precise right angles, adjust the chalk line so they will. Lift the line and let it snap.
4. With both working lines in proper position, lay loose tile from the center point along one line to the wall. Remember to allow for the space between tiles that will be grouted. If the space between the last full tile and the wall is less than one-half tile wide, adjust the original center working line one-half tile closer to the opposite wall. Snap the line. Repeat this process for the other line. Adjusting the lines in this manner ensures that the border tiles around the perimeter of the room will appear to be balanced.

The above method for establishing working lines will create a square pattern. Another popular option is a diagonal pattern. With this pattern the tiles are laid at 45-degree angles to the walls rather than parallel with them as they are in a square pattern. Although this method generally results in more tile cutting, the effect can be quite stunning. Establishing working lines for a diagonal pattern can be done in the following way:

1. Establish working lines as described in the steps above.
2. Measure and mark each line 4 feet from the point where the two lines intersect.
3. Tie a pencil to a string. Measure 4 feet from the pencil and cut the string. Scribe intersecting arcs for each of the four points you marked in step two by placing the end of the string on the point and drawing the arcs with the pencil.
4. After four sets of intersecting arcs have been made, snap chalk lines between the points where arcs intersect. The lines should pass directly through the center of original working lines. This produces four 90-degree angles at a 45-degree angle to the walls.

Whether you desire a square pattern or a diagonal pattern, you should always make sure your working lines are exactly perpendicular to one another. This

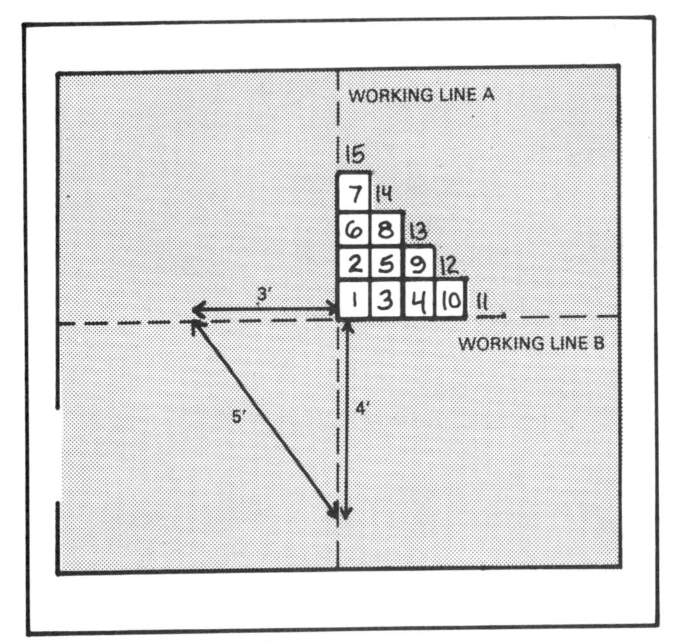

1. *Measure working lines.*

2. *Snap chalk line between points.*

3. *Lay loose tiles and readjust lines.*

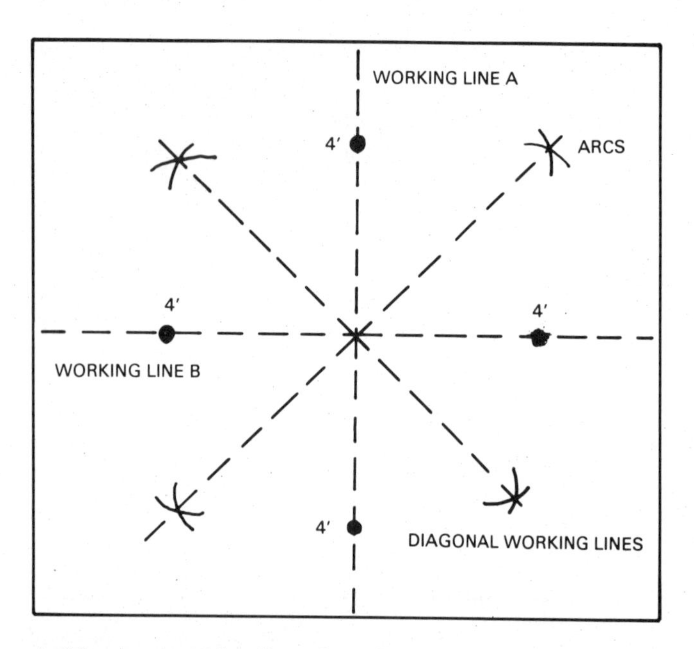

4. Measure working lines for a diagonal pattern if desired.

5. Snap chalk line between diagonal points.

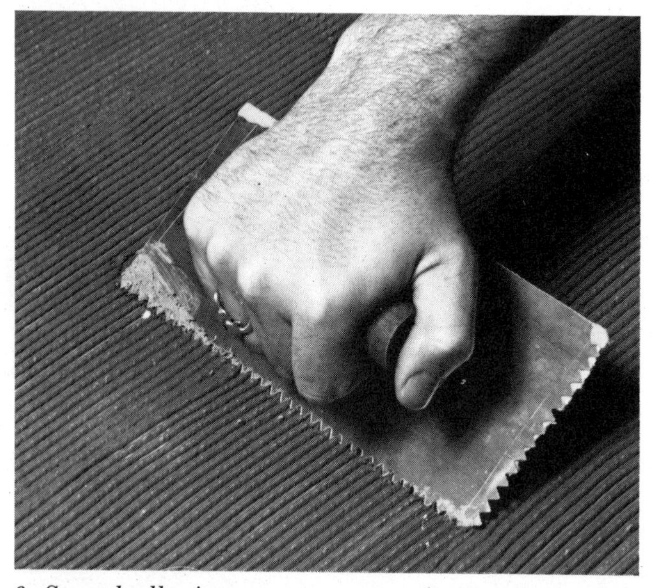

6. Spread adhesive.

is easily done. Measure and mark a point on one line 3 feet from the center point and then on the other line 4 feet from the same point; measure diagonally between these two points. The distance should measure precisely 5 feet.

Setting the Tiles

Before beginning the actual tile installation, make sure you have an adequate supply of materials and that all tiles are in acceptable condition. Mask off all areas that may be damaged by adhesive or grout with masking tape. The steps provided below describe the basic installation procedure for ceramic tile. Read carefully all instructions, directions, and product information. Actual installation procedures may differ with each manufacturer.

1. Following the instructions on the container label, mix the adhesive properly. Mix no more adhesive than you can use in about 20 to 30 minutes. Again, this open time may vary with some adhesives, so be sure to refer to the manufacturer's recommendations. Although some adhesives require different methods of application, the most universal method is with a notched trowel. Hold the trowel at about a 60-degree angle. Comb the adhesive over the area with the notched edge of the trowel to form ridges of adhesive and pull off excess adhesive. Do not apply the adhesive too thick or it will then ooze up between the tiles. Too little adhesive results in poor adhesion. Apply the adhesive up to, but not beyond, one set of working lines.
2. Begin setting whole tiles first. Tiles are laid in a pyramid sequence beginning in one of the quadrants formed by the intersection of the two working lines. Careful, accurate placement of these first tiles is essential for a professional-looking job because all other tiles are placed off them. Set the tile carefully into position with a gentle twisting motion, aligning its edges with adjacent tiles. Never slide a tile into position. Sliding causes adhesive to ooze through the joints. When the tile is in proper position, press it firmly into place with enough pressure to ensure a good bond between adhesive and tile. If you want a specific joint width, use uniform spacers between tiles. Remove spacers before the tile sets completely. Some ceramic tile is manufactured with built-in spacers.
3. Continue laying whole tiles. As you work, remove all excess adhesive from the tile before it

7. Lay tiles in a pyramid pattern. Do not slide tiles into position.

sets, using a cloth dampened with warm soapy water. Once set, adhesive is extremely difficult, if not impossible, to remove. Clean tools and hands as soon as time permits.

4. After finishing a quadrant, examine the area closely to make sure all lines and surfaces are level and uniform. Minor adjustments, if necessary, can be made while the adhesive is still tacky. Do not walk on the tile while the adhesive is drying.
5. Set the remaining quadrants in the same manner until all whole tiles are in place.

Cutting Ceramic Tile

When all full tiles are set, measure, cut, and place the remaining tile and trim. Tile can be cut one of several ways, depending on the tools that are available to you.

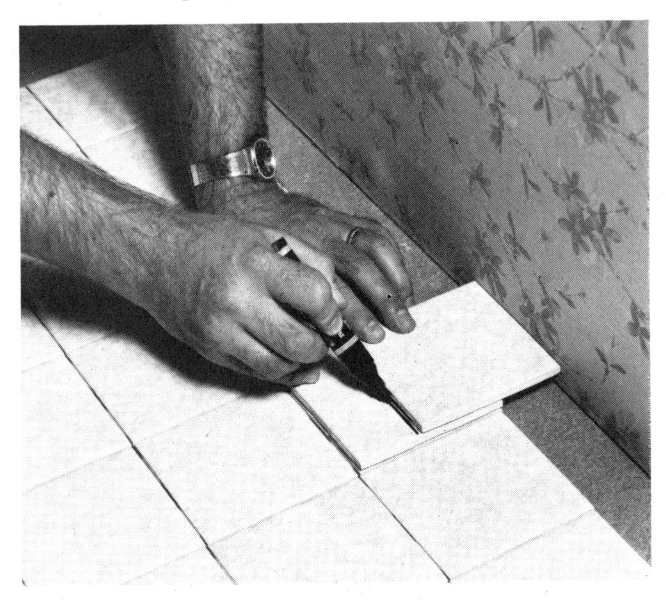

9. Mark border tiles for cutting.

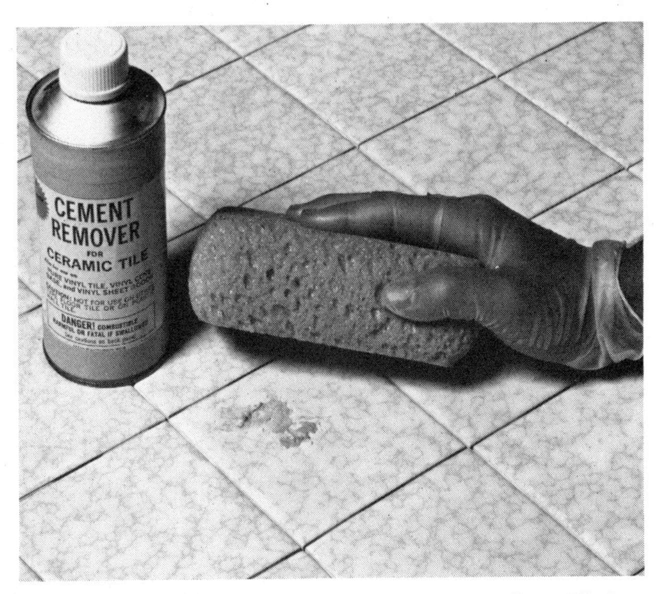

8. As you proceed, some of the adhesive may be spilled on the tile surface. Remove with recommended material.

1. Measure all border tiles and other tiles that require cutting. Border tiles can be measured for an exact fit by placing loose tiles directly on top of the last full tiles in the rows nearest the walls and/or obstacles.
2. On top of these tiles, place a third set of tiles. Slide these tiles until one edge touches the wall.
3. Draw a pencil line on the first loose tile by using the edge of the top loose tile as a guide. When you score these lines for cutting either with a tile cutter or glass cutter, score slightly inside the pencil lines to allow room between the wall and tile edge. This space serves as an expansion joint and can be filled with grout. This space will be covered by trim.
4. For curved or irregular cuts, like those required for tiles to fit around bathtubs or molding, use a contour gauge for marking tiles. Adjust the gauge to follow the contours of the shape. Transfer the pattern to the tile by following the contour gauge with a pencil.

Some tile dealers may cut marked tiles for you as a customer service or for a fee. Other dealers will rent or loan the tools necessary for cutting ceramic tile. It is even possible to cut tile with common household tools that you may have. Cutting tile is not difficult. Wear safety goggles when cutting tile. Many manufacturers make ceramic tile with ridged backs. If the tile you plan to cut has such a back, make the cut parallel to the ridges. The three basic ways to cut tile are with a commercial tile cutter, a glass cutter or a tile nippers. A masonry drill can be used to drill holes through ceramic tile.

The Tile Cutter This tool is made specifically for

10. For irregular shapes use a contour gauge to make a pattern of the contour.

11. Transfer pattern to tile by following contour gauge with a felt-tip pen or pencil.

12. Straight cuts are made on a tile cutter.

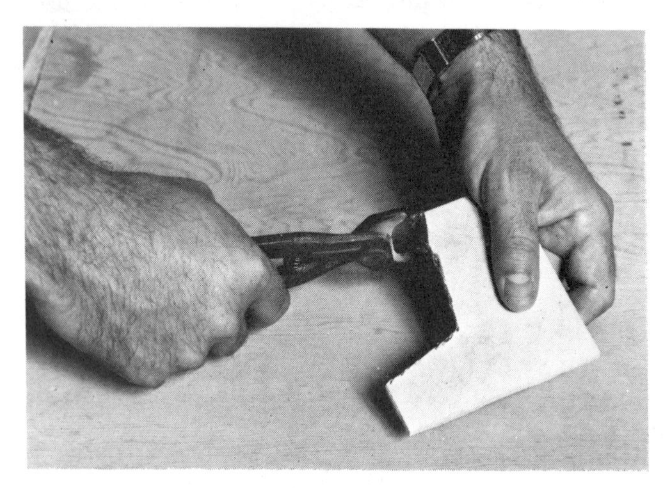

13. Use a tile nippers for irregular cuts.

cutting ceramic tile. It works effectively only on straight cuts.

1. Place tile on the cutter, face-side up.
2. Set angle gauge to size of desired cut by positioning the cutting wheel slightly inside the pencil line. This leaves room between the wall and tile edge.
3. Lift up on the tile cutter handle until a slight pressure is exerted on the tile surface. Draw the cutter across the face of the tile only once with a slow, steady pressure. Do not allow the cutting wheel to go off the edge of the tile or damage to the wheel may result. The cutting wheel scores the tile.
4. Press the breaking wings of the tile cutter near the edge of the tile or on the edge next to the angle gauge. Press down on the tile cutter handle with steady pressure to break the tile along the scored line.

Tile Nippers For irregular contours such as around molding or small circular cuts such as for pipe outlets, a tile nippers works well. A tile nippers allows you to nip off chunks of tile until you achieve the desired shape. Scoring the tile with a glass cutter first, if possible, will generally make a cleaner cut. Do not try to break off big chunks as this will damage the tile. As you get closer to the desired cut, test the fit every so often to prevent removing too much tile. If you only have a couple of tiles to cut in this manner, you can use a common pliers.

Glass Cutter If a common tile cutter is not available to you, straight cuts can be made with a glass-cutting tool. Place a straightedge along the pencil line on the tile and draw the tool along the edge of the tile to score a cutting line. Place the score line over a dowel or pencil and press down evenly on both sides. The

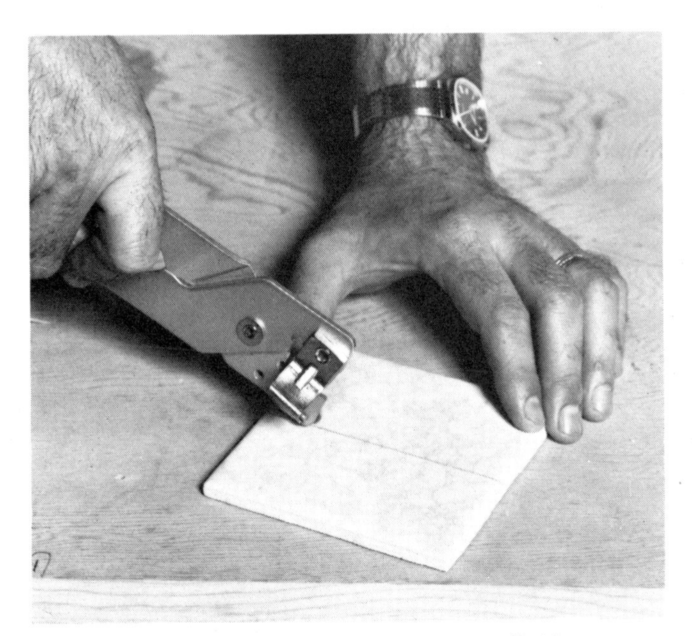

14. If a commercial tile cutter is unavailable to you, straight cuts can be made by scoring the tile with a glass cutter.

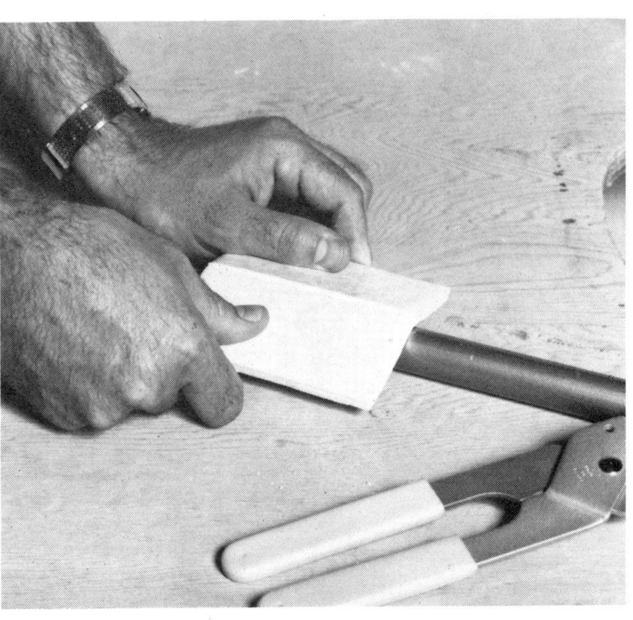

15. Position the score mark over a dowel or pipe and apply even pressure to both sides of the tile to break it.

16. A masonry drill can be used to drill a hole in any type of ceramic tile.

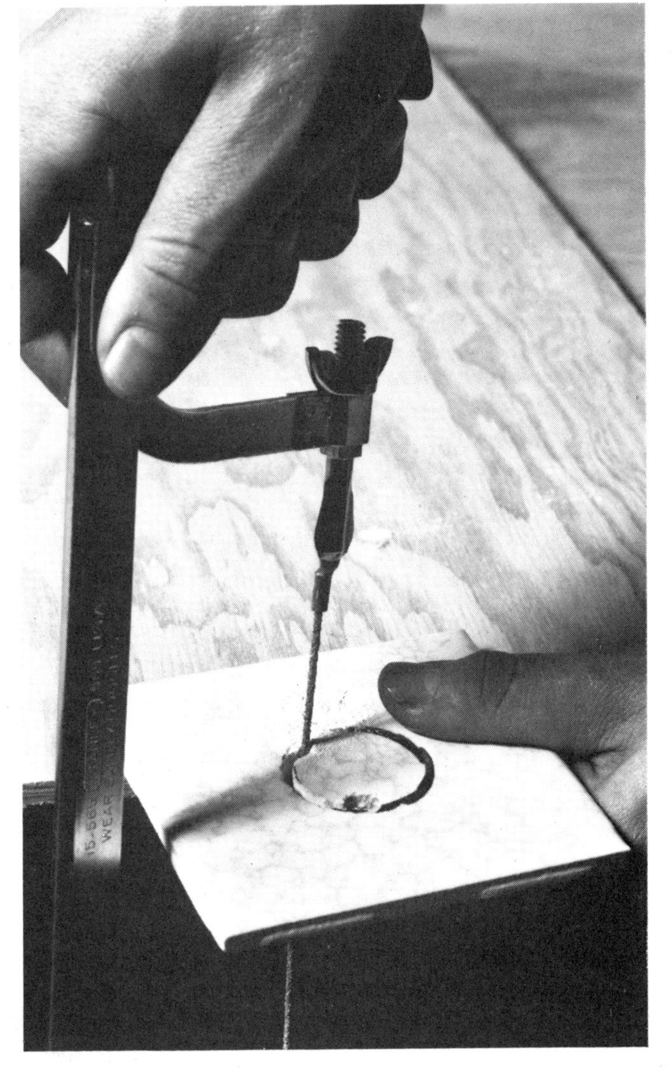

17. After drilling a hole, a larger hole can be made by using a rod saw.

18. Smooth all cut edges.

tile should break along the line. If you use a glass cutter to cut ceramic tile, it cannot be used to cut glass again.

Rod Saw A rod saw can be used for large irregular cuts. Use as you would any other handsaw. The rod saw is good for cutting pipe outlet openings.

Masonry Drill Sometimes cutting is not necessary. Small pipe openings or holes can be made easily and efficiently using a masonry drill bit. The drill makes a clean hole without having to cut the tile in half to create a round opening as you would have to do if using any of the above cutting methods. If the hole made by the drill is not quite large enough, it can be expanded with the rod saw without having to cut the tile.

After all cuts have been made, the rough edges of cut tile should be smoothed by rubbing them on an abrasive block or stone, a concrete block, or brick. The edges do not have to be perfectly smooth because grout will cover the joint and hide slight irregularities. Border tiles are now ready to be laid. Border tiles can be set using the method used for full tiles.

When all tiles have been set make sure all lines are straight and surfaces level. If minor adjustments are necessary, make them while the adhesive is still wet. Some large manufacturers recommend leveling and setting the tiles by tapping a large carpet-faced board with a hammer. Check the manufacturer's instructions. Make sure the entire surface is free of any adhesive.

Because ungrouted tiles break easily, avoid walking on the tile floor until the adhesive is thoroughly set, usually in 16 to 24 hours. Epoxy adhesives usually set fastest. If walking on the surface cannot be avoided, walk on a half sheet of plywood to distribute your weight evenly over several tiles.

Mixing and Applying Grout

When the adhesive has had adequate time to dry, it is time to apply grout. Clean all joints of excess adhesive. Grout fills the joints between tiles, bonding them together for greater structural strength. Grout also has a visual function. Since grouts are available in several colors or can be changed to a particular color using pigments, you can select one that complements or contrasts your particular tile design or color. It can be used to unify a color scheme or to create interesting accents that enhance overall decor.

Mix grout carefully according to the manufacturer's directions. Most grouts form a paste the consistency of peanut butter when mixed. Mix only what you can use in 10-15 minutes, or it will not work properly. Remix after several minutes. Wear rubber gloves when mixing or applying grout to prevent skin irritation.

Most grouts are applied in the same way except silicone rubber grouts which are applied with a caulking gun. Other grouts are applied in the following manner: fill the joints, remove excess, clean the tile, and allow to cure.

1. Apply grout to the tile surface with a grout float or rubber-faced float. Work it into the spaces between tiles, forcing it to completely fill all joints. Go over each area several times to ensure that the grout is thoroughly worked into the joints. To increase the working life of cement grouts, moisten the tiles slightly before application. This prevents the ceramic tile from absorbing moisture from the grout.
2. Wipe off excess grout by drawing the edge of the float diagonally across the grout lines. Remove as much excess as possible.

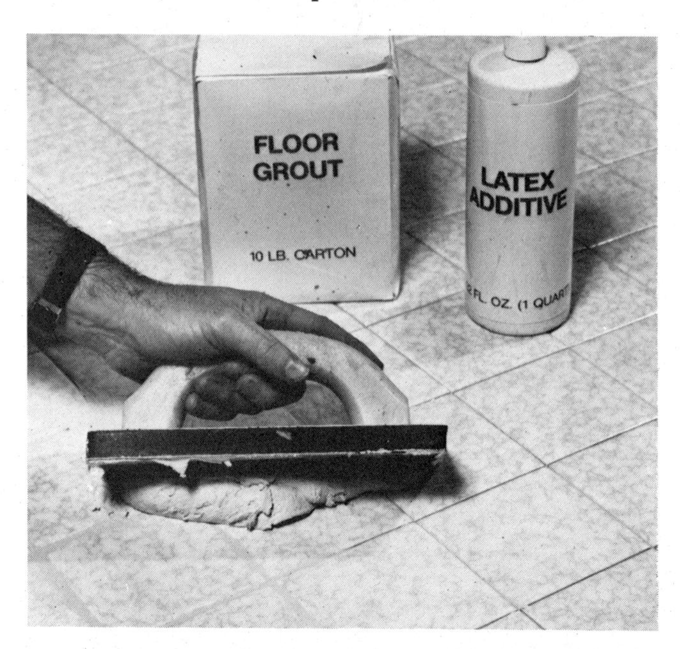

19. Apply grout with a rubber-faced float.

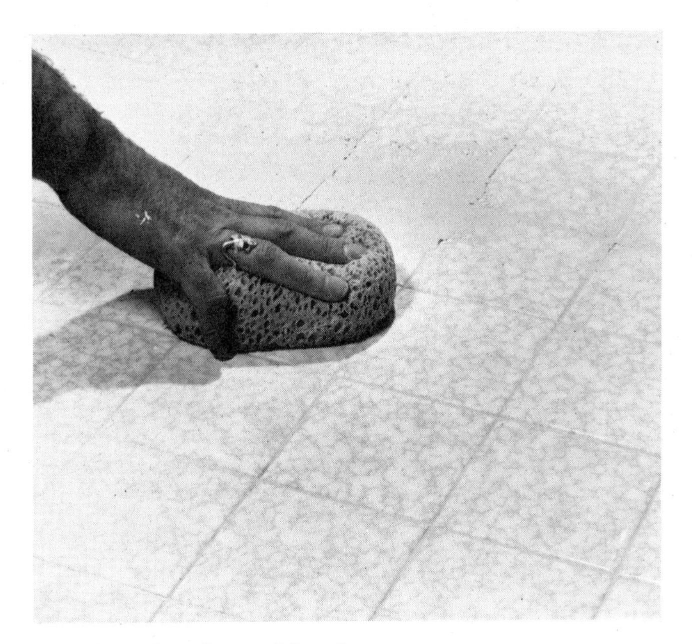

20. Wipe tiles clean with a damp sponge.

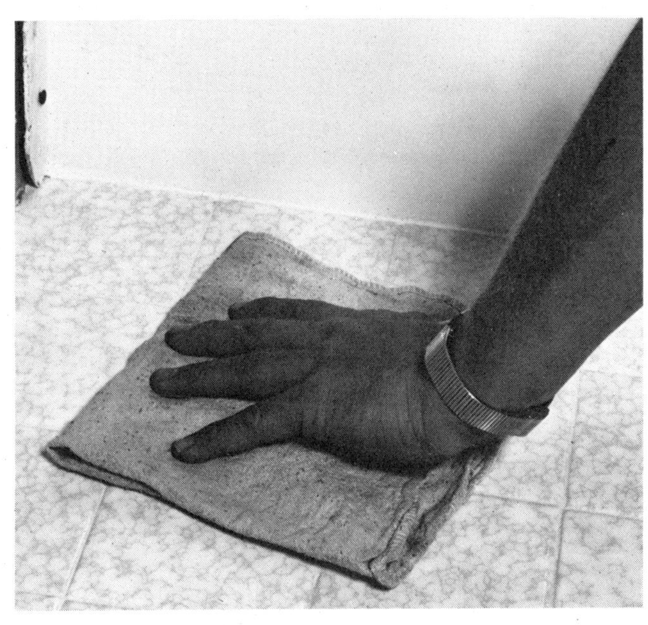

21. Polish dried surface with soft abrasive cloth.

3. Wait approximately 15 minutes, then carefully wipe the tiles with a clean, damp sponge. Do not allow the grout to dry before cleaning; it cannot be removed. Use a circular motion with light even pressure to avoid damaging the fresh grout. Rinse the sponge in clean water frequently. The sponge creates a slightly rough surface texture. If you desire a smoother finish, use a jointer to make smooth joints.
4. After the grout has dried, the tiles will probably have a hazy film on the surface. This can be removed by polishing with a dry soft abrasive cleaning pad or cloth. Allow the grout to cure for at least 72 hours. If possible, do not walk on the surface until the grout is cured. If foot-traffic is unavoidable, walk only on plywood placed over the tile.

Dry Grout Process

An alternative method to applying grout in the manner described above is called the dry grout process. This method of application can also be used on mosaic or quarry tile floors. The chief advantages of this process are that it cuts grouting time by 2 to 3 hours and produces smoother grout lines. The process also ensures more even color, helps prevent washed out grout lines, and, because of the dry grout's abrasive nature, cleans the tile surface in the application process.

The tile must be firmly set before applying grout. Try to allow 24 hours between setting tile and grouting. The dry grout process can be used only with sanded grout, which is available premixed with additives or additives can be purchased separately and mixed at the site.

1. Mix the grout according to the manufacturer's instructions. Mix only enough at one time that can be used in 10 to 15 minutes. The mixture should form a paste the consistency of creamy peanut butter. Allow it to stand for approximately 10 minutes and re-mix. It is wise to wear rubber gloves to protect hands when mixing and applying grout as the additives can cause skin irritation.

2. Apply the grout, working in approximate 3 x 4-foot sections with a rubber-faced float. Spread the grout over the tiles, forcing it into the joints. Work the grout into the joints by going over the area 2 or 3 times with the float.

3. Holding the float at a 45-degree angle to the tiles, wipe off the excess grout by drawing the edge of the float diagonally across the grout lines.

4. The excess grout film that remains can be wiped away with a damp sponge.

5. After wiping the tile surface clean, allow the grout to dry approximately 10 minutes. Lightly sprinkle several handfuls of dry powdered grout over the grouted tile surface. Vigorously buff the tile with a soft terry cloth towel. Shake the cloth frequently to remove excess powdered grout. Continue this buffing action until the surface of the tile is clean and the joints are full and smooth. If the grout lines appear ragged on the edges, wipe down with a damp sponge and repeat this step.

6. Lightly brush or sweep any excess dry grout from the tile surface. Dispose of this material; it should not be reused.

1. Apply grout with float.

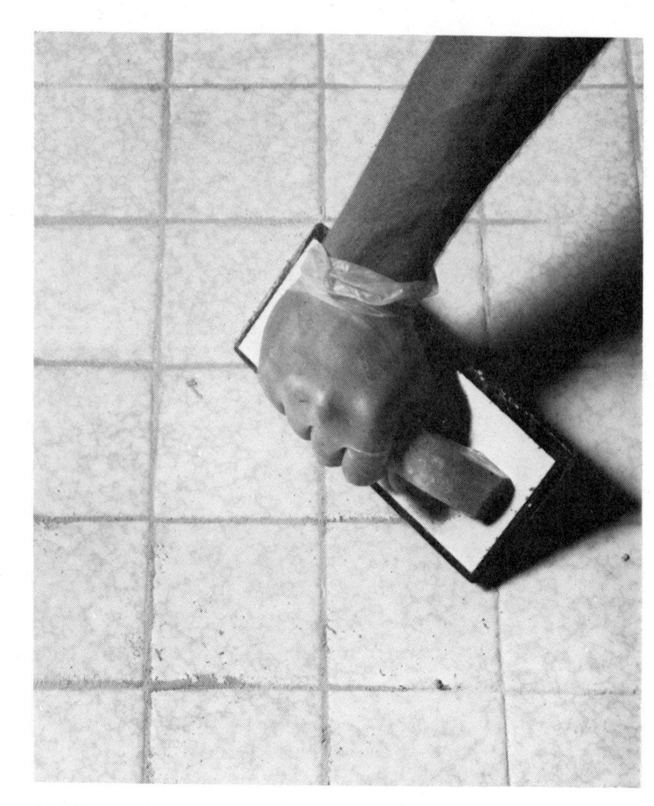

2. Wipe off excess grout with float.

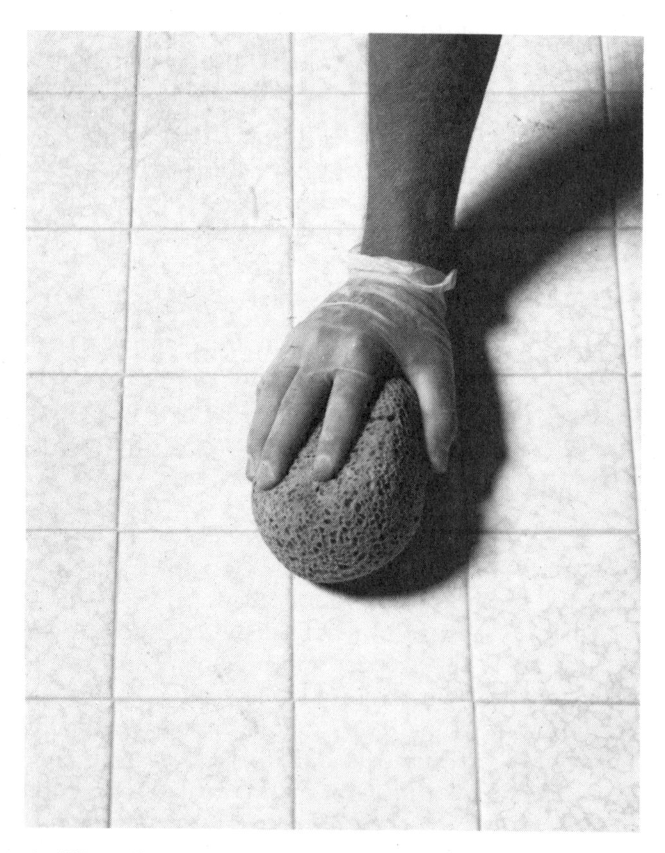

3. Wipe tiles clean with damp sponge.

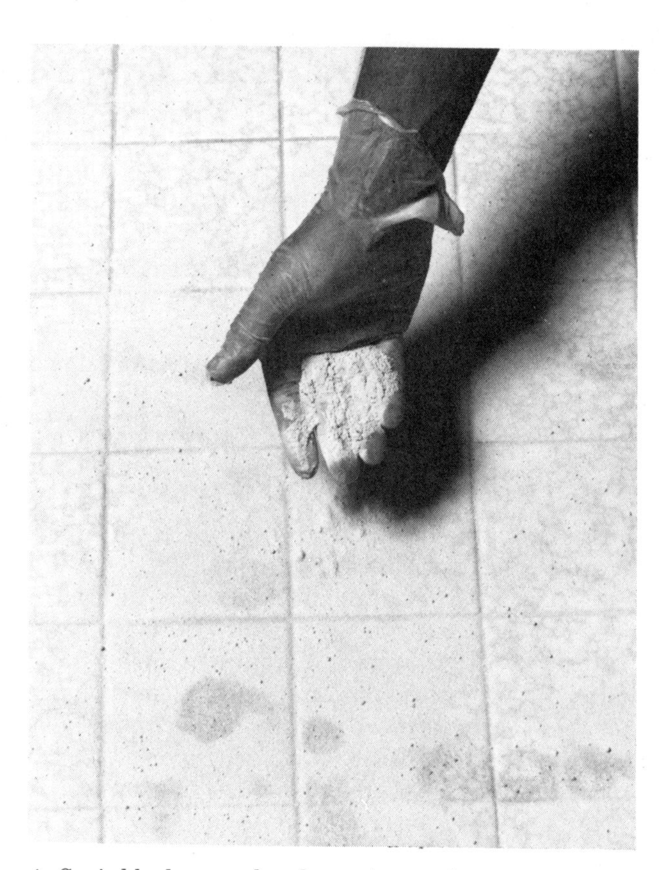

4. Sprinkle dry powdered grout on surface.

5. *Vigorously buff the tile with towel.*

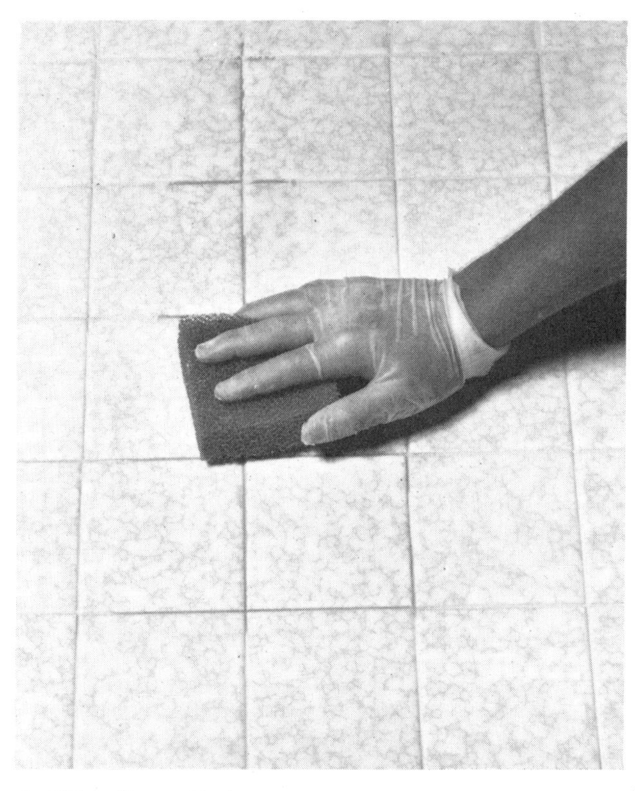

6. *Wipe tiles with damp sponge.*

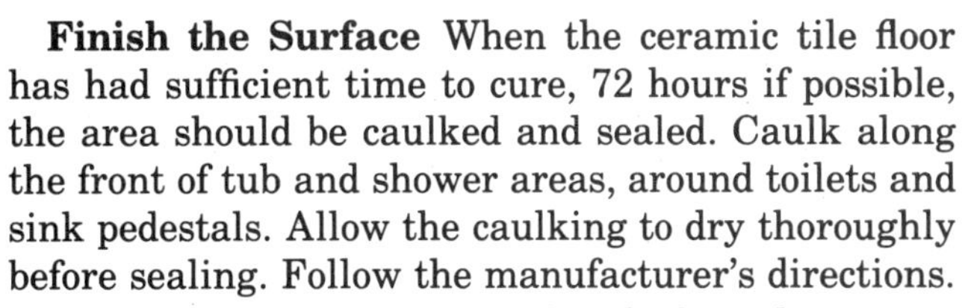

Finish the Surface When the ceramic tile floor has had sufficient time to cure, 72 hours if possible, the area should be caulked and sealed. Caulk along the front of tub and shower areas, around toilets and sink pedestals. Allow the caulking to dry thoroughly before sealing. Follow the manufacturer's directions.

It is not necessary to seal glazed tile and epoxy or silicone rubber grouts. Most sealers are silicone based. The sealer penetrates unglazed tile and mortars and prevents these materials from absorbing moisture. Sealers wipe off glazed surfaces easily. Many sealers are available with spray applicators. Apply according to the manufacturer's directions. Do not apply too much sealer. Buff the surface with a clean towel when the sealer is dry. This removes any film and imparts a rich luster to the tile.

Replace all moldings and the ceramic tile floor is ready for use.

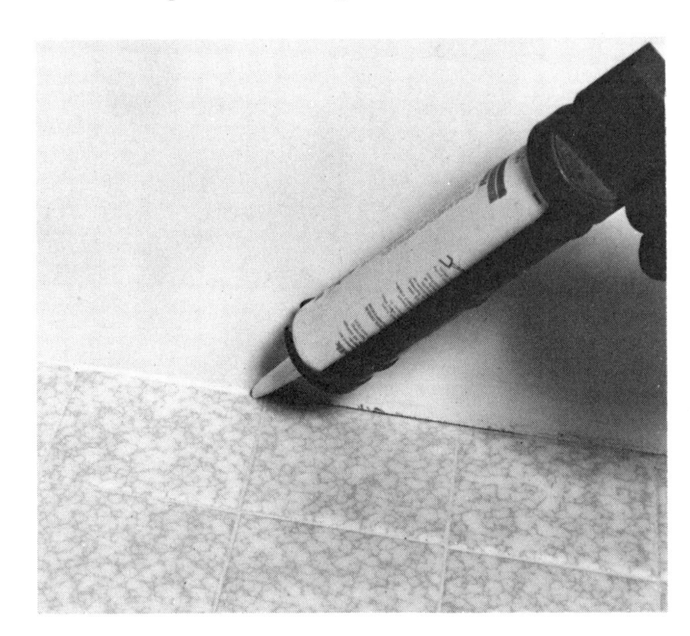

22. *Caulk around walls and obstacles.*

23. *Seal the dried tile surface.*

Installing Quarry Tile

Quarry tile can add the perfect touch of sophistication to your home and offers a lifetime floorcovering to any room. Its durable composition can handle heavy traffic areas. Preparation and installation techniques are, for the most part, very similar to those for other types of ceramic tile as described in the last section. For more detailed instructions, please refer to the information presented in the "Installing Ceramic Tile" section.

1. Prepare the subfloor surface as you would for ceramic tile according to the type of adhesive you will be using and the manufacturer's directions for that adhesive.
2. Establish working lines as you would for ceramic tile.
3. Most of the tools and supplies used for ceramic tile installation can be used for quarry tile installation as well.

Setting Quarry Tile Assemble all tools and supplies. Make sure all cartons of tiles are uniform and tiles are in acceptable condition.

1. Mix the adhesive according to the manufacturer's directions. Apply the adhesive to the subfloor with a notched trowel. Work only on one quadrant at a time.
2. Set all whole tiles first, using the pyramid sequence described in the last chapter. Quarry tiles generally have wider joints than ceramic tiles. To achieve wider even joints, you can do one of two things: add spacers between each tile to achieve uniform spacing (many tile manufacturers provide plastic spacers), or use a line

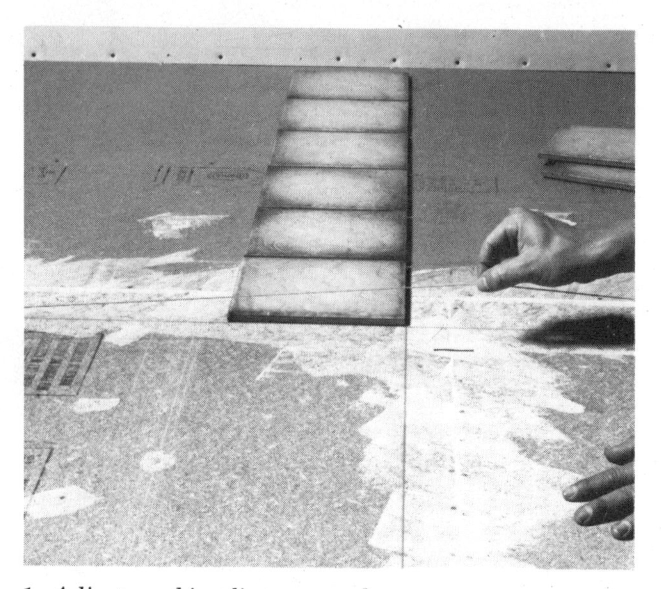

1. Adjust working lines properly.

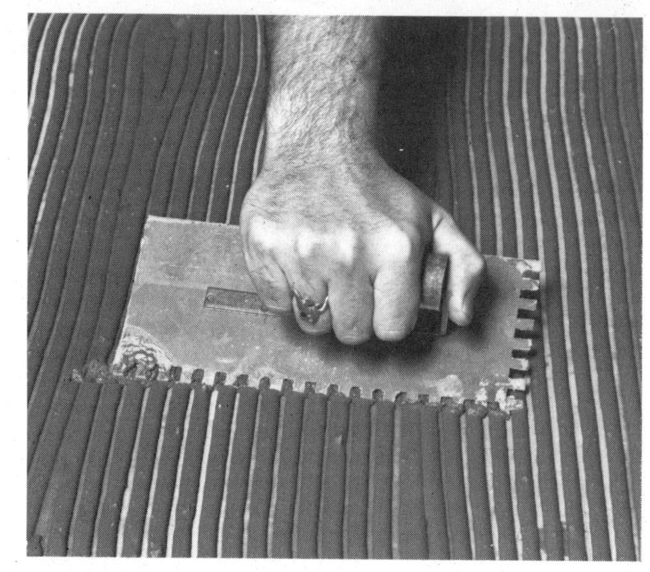

2. Spread adhesive on one quadrant.

3. Lay tiles in pyramid sequence. White spacers ensure uniform spacing.

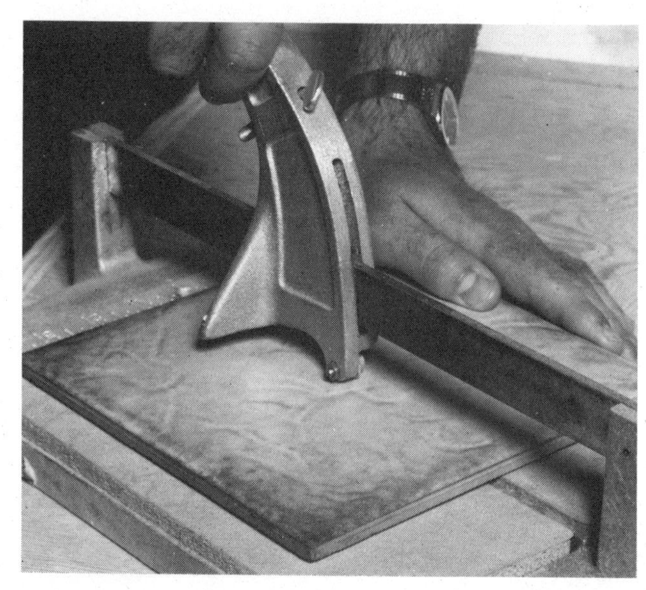

4. Mark and cut all border tiles.

Ceramic Tile Floorcoverings

The natural appearance of this modern open kitchen design is established primarily by the basic quarry tile floor. The rich earth tone of the tile blends well with all design elements.

Ceramic floor tile is not limited to use only on floors. Tub enclosures and countertops can also be covered with ceramic tile (left and above). An entire bathroom covered with tile is easy to clean and naturally beautiful.

The dark quarry tile contrasts well with white grout and bright blue fixtures to make a compact but stunning bathroom (below).

Mosaic tile (above) blends well with white ceramic tile and patterned mosaic wall tile in this well-designed bathroom.

Mosaic and ceramic tile (right) ties together all other design elements in this small attractive bathroom. Some of the floor tiles are used as a wall accent.

Mosaic tile combines with cedar planks (above) to create a clean, rough appearance in this contemporary bathroom. The natural materials require little maintenance.

Quarry tile helps create a rustic-looking kitchen (left). It balances nicely with the texture of the stone and rich tones of the oak cabinetry.

Quarry tile in three different colors provides an attractive, easy-to-clean floor for this foyer (right).

Resilient Floorcoverings

Resilient floorcovering is becoming increasingly popular on family room and living room floors (above) because of its wide variety of colors and patterns.

Resilient floor tiles are easy to install and offer a realistic imitation of more expensive floors such as wood parquet (right).

Resilient tiles (left) enhance wallcoverings, curtains, framed artwork, and furnishings in this open kitchen/dining area.

The light tones of this resilient floor (above) reflect sunlight well to help establish the bright, clean look of this kitchen.

The subtle pattern of these resilient tiles (above) is intensified by the plants, cabinet design, and accent furnishings in this kitchen.

Resilient floorcovering provides a durable, easy-to-clean surface for high-traffic areas like entryways (below).

The entire design scheme of this kitchen/dining area is tied together by the rich colors of the sheet resilient flooring. The flooring should set the mood of a room.

Wood Floorcoverings

Hardwood strip or plank floorcoverings are perfect for attic conversions. In this children's bedroom/ play area (above) the strip flooring picks up the same pattern created on the walls.

Teak parquet flooring (right) creates a beautiful entryway that can set the tone for the entire interior design scheme.

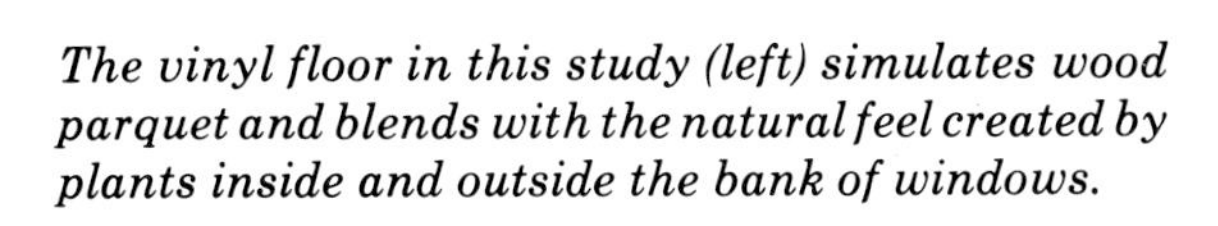

The vinyl floor in this study (left) simulates wood parquet and blends with the natural feel created by plants inside and outside the bank of windows.

Oak parquet flooring (left) is the base for this elegant formal dining room. It enhances the fine wood furniture very nicely.

Traditional wood parquet (right) is as easy to install as resilient tiles and provides a rich, elegant look in this living room.

Wood flooring is as much at home in a rustic one-room vacation home (below) as it is in a formal dining room.

Carpeting

New fibers and designs make carpeting suitable for traditionally wet areas like bathrooms (above). Here the rich brown carpeting highlights the wood vanity and cabinets.

Luxurious shag carpeting creates a comfortable informal sitting area in front of this massive fireplace (right).

Nothing can compare to the comfort and warmth offered by carpeting, particularly in bedrooms (above).

In children's bedrooms, carpeting provides a soft comfortable surface that gives protection against falls and roughhousing (below).

gauge which automatically measures and establishes spacing between quarry tiles. Make sure all lines and joints are straight and uniform.

3. Remove excess adhesive from tile while adhesive is still wet.
4. Set the remaining quadrants in the same manner.
5. Measure and cut quarry border tiles as you would ceramic border tiles.
6. Allow the adhesive to dry 24 hours before grouting.
7. For unglazed quarry tile use a recommended grout release.
8. Mix grout according to the manufacturer's directions.
9. Apply grout to tile surface with a rubber-faced trowel. Work thoroughly into joints.
10. Sponge excess grout off tiles with a damp sponge.
11. Remove film by buffing with a dry, soft cleaning pad.
12. Caulk and seal the grouted surface after the grout is completely cured, usually 72 hours to 2 weeks.
13. Buff sealed surface with a clean rag.

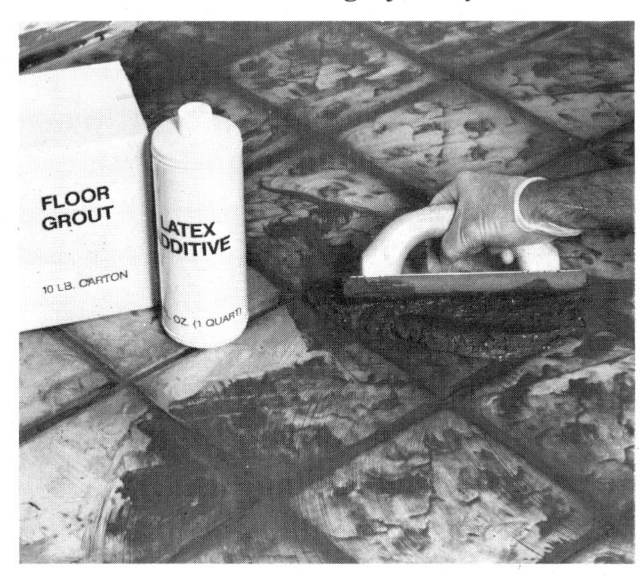

5. Apply grout with rubber-faced trowel.

6. Sponge off excess grout.

7. Caulk around edges and obstacles.

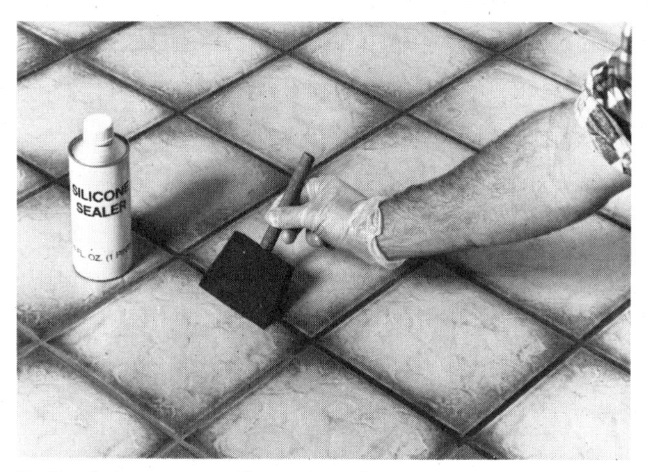

8. Seal the quarry tile surface.

9. Buff surface with a clean cloth.

Installing Mosaic Tile

Mosaic tile is easy to install and creates a tough, durable finished floor. Since many of the preparation and installation techniques are the same as those for ceramic or quarry tile, read the section on "Installing Ceramic Tile." This information will make the installation of a mosaic tile floor faster and easier. Since most mosaic tiles are available in sheet or panel form, installation is already easier because a group of tiles are prespaced and held together mechanically on a backing sheet. Some panels are even available pregrouted.

1. Prepare the subfloor surface as you would for ceramic tile. Follow the manufacturer's directions for subfloor preparation on the container of the adhesive you plan to use.
2. Establish working lines as you would for any ceramic tile floor.
3. Most of the tools and supplies used for ceramic or quarry tile floors can be used for mosaic tile installation.

Setting Mosaic Tile Assemble all tools and supplies. Make sure all sheets or panels are uniform and in acceptable condition. All tiles should be square with each other on the backing sheet.

1. Mix the adhesive according to the manufacturer's directions. Apply the adhesive to the subfloor with the proper notched trowel. Since sheets of tile are quite a bit larger than single ceramic or quarry tiles, you can apply adhesive to a larger area.
2. Begin setting whole sheets of mosaic tile at the intersection of the two working lines. You will be working on two quadrants at the same time, alternating sides as each sheet is laid. Set all full sheets first. Leave the same amount of space between sheets of mosaic tile as there is between individual tiles on a sheet. Many mosaic tile sheets have distinct patterns or designs. Position such sheets so the patterns are consistent for the entire area.
3. Remove excess adhesive from all tiles.
4. Set the two remaining quadrants in the same manner.
5. Measure and cut mosaic border tiles as you would ceramic border tiles. Remove individual tiles that must be cut from the backing sheet. Cut away excess backing from the area of the tile sheet that will not be used.
6. Allow the adhesive to dry 24 hours before grouting.
7. Mix and apply grout according to the manufacturer's directions.
8. Caulk and seal the grouted surface when cured.
9. Buff sealed surface with a clean rag.

1. Adjust the working lines after laying loose sheets of tile.

2. Spread the adhesive with a notched trowel.

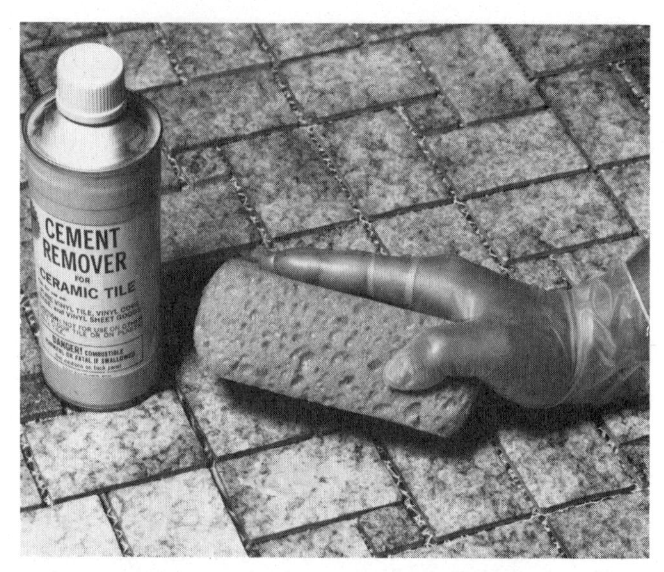

3. Use cement remover to clean adhesive from the mosaic tile surface.

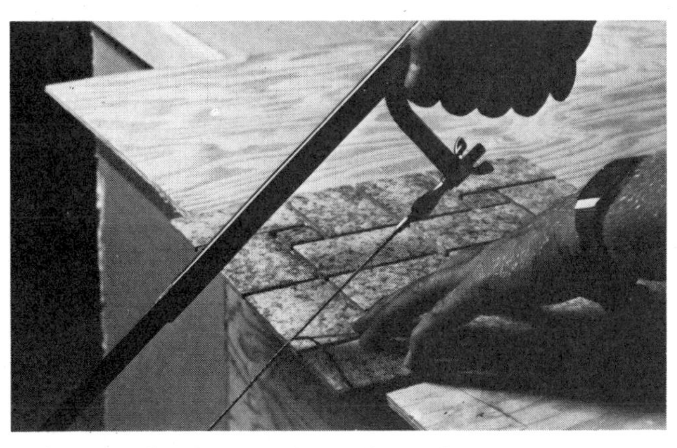

4. Saw marked border tiles with a rod saw.

5. Trowel on grout, working it into all cracks.

6. Seal the mosaic tile surface.

7. Buff surface with a clean cloth.

Caring for Ceramic Tile

A ceramic tile floor, whether it is ceramic, quarry, or mosaic tile, is one of impressive natural beauty. Actually, ceramic tile is so durable that very little care and maintenance are required to keep the floor looking as it did at the time of installation. The sealer you applied to the finished floor protects the grout and unglazed tiles from water and moisture.

Periodic cleaning is all that is necessary to keep a ceramic tile floor looking like new. Sweep or dust mop the area to remove loose dirt. Wash tile with hot water and a mild detergent or cleaner; rinse thoroughly and dry with a soft rag. Several tile manufacturers have commercial grout and tile cleaners. Follow the manufacturer's directions when using these preparations. Commercial household bleach applied with a toothbrush or other soft brush will clean grout very well. Water and mineral deposits can be avoided by wiping tile dry with a soft cloth when wet. Wax can be applied over a cleaned ceramic tile floor, provided the floor area is not exposed to water. Follow the manufacturer's directions when applying wax.

Ceramic tile can become stained despite its hardness and durability. The following chart identifies common stains and the recommended method of removal.

Stain	Cleaning Method
Alcohol, coffee, fruit juice, blood, animal stains, tea, ink, food	If conventional cleaning methods do not effectively remove stain, use scouring powder and a brush. Household bleach may lift the stain if allowed to remain on the stain a few minutes before rinsing. Do not use bleach on colored grout.
Tar, grease, oil, lipstick, heel marks	If commercial tile cleaners do not work, dampen a soft cloth with mineral spirits and rub lightly. Clean the surface thoroughly after removing the stain with mineral spirits. Ammonia or lighter fluid can be substituted for mineral spirits.
Paint	If the stain is relatively old, scrape as much of it off as possible with a razor blade, taking care not to scratch the surface. Apply paint remover, let stand, rub with brush, and clean thoroughly.

1. Scrape out grout from around the damaged tile.

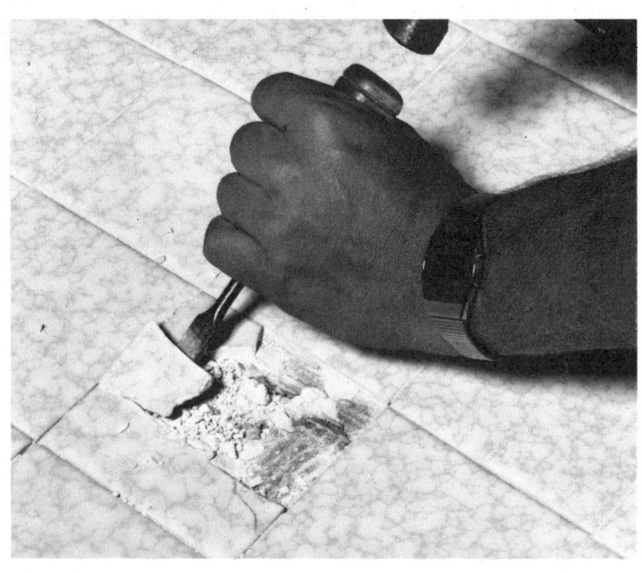

2. Chip out damaged tile.

3. When all tile fragments are removed, apply adhesive to new tile and insert.

Repairing a Broken or Damaged Tile

Ceramic tile can become chipped, cracked, or loose with repeated use. The grout can become stained or can loosen in the joints. Below are listed several common problems and their respective repairs.

Damaged Grout Grout that has become worn, cracked or badly stained can be replaced. Scrape out the old grout from the joints with a screwdriver or narrow chisel. Clean joints well and apply grout as described in the "Installing Ceramic Tile" section.

Loose Tiles Clean all old adhesive and grout from the tile(s) and subfloor. Spread new adhesive on the tile(s) and drop into place. Regrout when the adhesive is completely dry.

Damaged Tiles Cracked, broken, and chipped tiles should be replaced for best results. It is not a difficult job, provided you use care and patience. Follow the step-by-step instructions listed below.

1. Scrape out all grout surrounding the damaged tile.
2. Score an X on the tile surface with a glass cutter.
3. Use a hammer and cold chisel to chip out the old tile, beginning at the center of the tile.
4. Remove adhesive and any remaining grout from the subfloor. Patch the subfloor if it is damaged.
5. Spread adhesive on back of new tile, keeping it about ½ inch from the tile edges.
6. Carefully drop tile into place and tap into adhesive using a hammer and wood block.
7. Fill joints with grout as described earlier in this chapter.

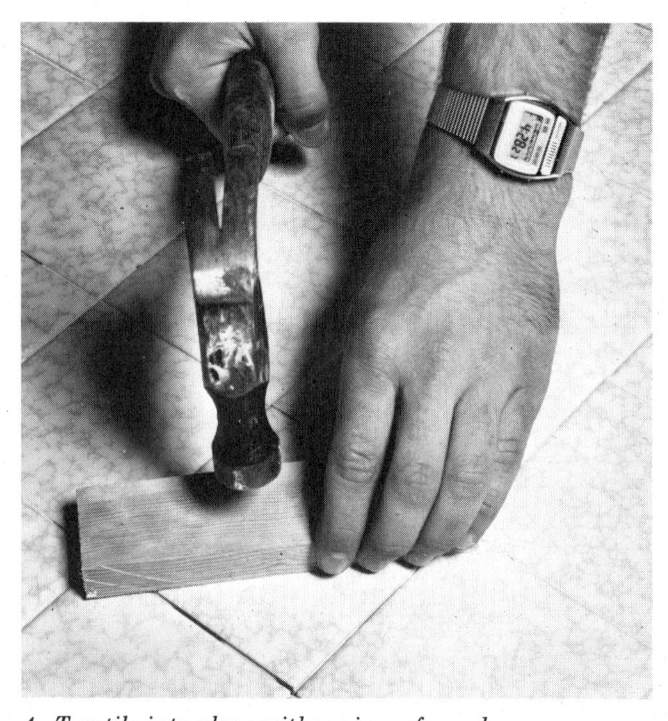

4. Tap tile into place with a piece of wood.

Resilient Floorcoverings

Resilient floorcoverings, more commonly referred to as vinyl floorcoverings, have experienced a steady increase in popularity over the years primarily due to the easy care characteristics and the wide range of designs, colors, and patterns. Most manufacturers have designed their products for easy do-it-yourself installation, even sheet goods which were once cumbersome and difficult to install properly. Vinyl floorcovering installation requires little skill and only a few basic hand tools. These new vinyl floorcoverings have all but replaced linoleum and asphalt and asbestos tile.

Resilient flooring provides the do-it-yourselfer with an opportunity to install a floor that is economical and beautiful. In fact, installing a resilient floorcovering yourself can save up to one-third the cost. Resilient flooring is already one of your best floorcovering buys, and when you can eliminate labor costs, your savings can be substantial.

Two Basic Types of Resilient Flooring

Resilient flooring is available in two basic forms: tiles and sheet goods. Tiles are individual squares most commonly 12 inches square. Sheet goods are continuous rolls that are generally 6 or 12 feet wide.

Tiles have long been the favored choice because they were easier to install and less expensive. The primary advantange of sheet flooring is few or no seams, creating a more finished, continuous appearance and allowing for more sophisticated design styling. In recent years, however, tile manufacturers have duplicated many of the designs of the most expensive sheet flooring and have even found a way to conceal the seams between individual tiles. On the other hand, manufacturers have made sheet goods almost as easy to install as tiles. The price differential is narrowing as well.

Tiles Asphalt and linoleum tiles were the most popular flooring materials for many years because they were inexpensive and fairly easy to install. They were hard to maintain and clean, however, did not retain their color well, and often cracked or chipped. Solid vinyl and vinyl composition tile have made earlier varieties obsolete.

Solid vinyl tile is composed of exactly what the name implies. The more vinyl in a floorcovering, the higher the cost; so this tile is the most expensive resilient tile. Solid vinyl tiles are more colorfast, offer greater resiliency, are more durable, and are easy to care for. Vinyl composition tile is composed of vinyl and another material, usually rubber. The addition of another material makes it harder and more suitable to high-traffic areas. This tile offers excellent resistance to moisture, dirt, and wear.

Tiles can be set in adhesive or may be purchased with a selfstick backing for even easier installation. Resilient tiles that are set in adhesive applied to the subfloor are called dryback tiles. Selfstick tiles have adhesive preapplied to the back side of the tile. Installation is simple: peel off the paper backing, position, and press into place. An entire room can be floored in less than half a day.

Many tiles require periodic waxing and buffing to protect the surface and maintain a high-gloss finish. One of the major resilient floorcovering advancements in recent years is the development of a no-wax surface. A durable urethane coating is bonded directly to the surface of the tile. This provides a strong surface that retains its shine without waxing and buffing, making care and maintenance much easier. A damp mopping will generally restore the original brilliance of the floor. There are several types of no-wax surfaces. Some require periodic polishing and others do not. Follow the manufacturer's maintenance recommendations in any case.

Sheet Goods Since resilient sheet flooring has few or no seams, it is perfectly suited for bathroom or kitchen floors—two rooms that are susceptible to water spills. Water can work its way into seams and cracks and loosen the adhesive bond. There is less chance of this happening when the floor is virtually seamless.

Working with a large sheet of flooring is more difficult than working with 12-inch square tiles, but it is easier now than ever before. At one time the sheets were stiff and unwieldy. They had to be laid quickly and carefully before the adhesive dried. If the flooring was not cut perfectly, a much harder task in the past, the job was either sloppy or a total loss. Fortunately, things have changed. Today sheet flooring is more flexible and lightweight. Some no longer require adhesive, only staples around the perimeter. Miscutting is perhaps the biggest problem, but several manufacturers now provide pattern paper for making a scale drawing of the room, a cutting tool, and detailed instructions for cutting and installing. Sheet flooring can be laid directly over a problem floor that would require resurfacing with underlayment for other types of floorcovering.

Sheet vinyls are produced in one of two ways: by an inlaid process or by rotogravure printing. The inlaid process is a tedious one that builds up colors and patterns from thousands of vinyl granules which are then fused together with intense heat and pressure. Floors manufactured in this way have excellent durability and deep, rich colors that have a crafted appearance. Rotogravure combines photography and printing. Almost anything that can be photographed can be printed with a high degree of realism. The printed design or pattern is protected by a core of foam cushioning and a top layer of clear vinyl. Inlaid vinyls are generally more expensive than the rotovinyls.

Choices and Selection

Resilient flooring, whether tile or sheet goods, offers an amazing variety of colors, patterns, textures, and designs that will fit comfortably into any room decor. Styles range from traditional to contemporary with new styles being introduced on a semiannual basis. In the past few years many manufacturers have been able to simulate a wide range of hard-surfaced materials with excellent accuracy. Sheet and tile resilient flooring now available closely resembles wood parquet, wood plank, ceramic tile, brick, or stone.

When selecting a resilient flooring, there are several other important factors aside from color, texture, and pattern to consider. First consider the wearability of the floorcovering. If the floor will be subject to heavy traffic or possible spills and stains you will need a very durable surface. If, however, the floor will be in a child's room, the child may outgrow the design in a few years and a less durable and less expensive resilient floorcovering should be considered. The best value in resilient floorcoverings offers the most suitable performance for the area in which it is installed.

Check into the floor's ability to resist stains and discoloration of the wear surface. Quality resilient floorcoverings should have a surface that is tough and dense enough to resist most household stains. The floor should keep stains on the surface where they can be easily wiped away. Also check into the ability of the floor to resist discoloration from excessive sunlight or concentrated heat. Most floors fade and discolor when exposed to periods of strong direct sunlight. If the room in which the floor will be installed has direct light sources, select a floorcovering with high fade-resistance. Concentrated heat from heat registers or baseboard heaters can also cause discoloration.

Another consideration, although a bit less important than the others, is comfort. Resilient floorcoverings are manufactured with varying amounts of cushioning. Some are very resilient and others are quite hard. For areas where long periods of standing are required or where children often play, select a floorcovering that offers good resiliency.

When you have decided on a particular type of resilient flooring, it is wise to make a scale drawing of

Resilient tiles laid in a diagonal pattern make an excellent floor for this hobby/recreation room.

the area to be covered. The method for drawing this planning aid is described in the chapter on estimating materials and making a scale drawing. Take it with you to the floorcovering dealer. Order extra materials for a built-in margin of error and for repair of damaged flooring in the future. Excess tiles can often be returned. Have the materials delivered or pick them up two or three days before you plan to install them, so they can become acclimated to the surrounding environmental conditions.

Where to Use Resilient Flooring

Resilient flooring can be used in any room of the home although it is most commonly used in bathrooms, kitchens, recreation rooms, and utility rooms. Since one of the main features of resilient tile is easy maintenance, it makes an excellent floorcovering for areas that are subject to dirt, water, and spills. A quality cushioned vinyl provides a relatively soft floor that is easy to clean, especially appropriate for a child's bedroom or play area. Resilient floorcoverings are probably the most versatile of all floors, suitable to any area, traffic pattern, or room decor.

Resilient sheet flooring can withstand the daily punishment an entryway floor must take.

Ease of maintenance makes resilient flooring an excellent choice for dining rooms.

Resilient tiles blend well with the wood cabinets, stone fireplace, and furniture in this basement recreation room.

Preparing the Subfloor

Subfloor requirements for resilient flooring are basically the same as for other types of floorcoverings. The subfloor surface must be level, clean, dry, and structurally sound. Since resilient floorcoverings are so pliable, they will conform to minor subfloor irregularities. If an absolutely smooth floor is not required, these irregularities do not have to be repaired.

Resilient flooring can be installed over concrete, plywood sheet subfloor, board subfloor, or even some existing floorcoverings. Before preparing the actual subfloor surface, check the supporting framework for any structural problems. Common repairs are described in the basic floor construction chapter. Also remove molding, taking care to mark the position of each piece, so the molding can be renailed in the proper position. Remove all other obstacles such as furniture, furnishings, and heat registers.

Concrete Subfloor

A concrete slab must be totally free of oil, grease, dirt, paint, and finish. If these are present, they must be removed with the proper solvent. The most important requirement, however, is that the concrete slab must be completely dry. Moisture will cause resilient flooring to loosen. The moisture content of concrete can be tested in one of several ways. These simple tests are explained in the subfloor preparation section of the wood flooring chapter.

If the concrete subfloor is clean and dry, check for any low spots or cracks. These are easily repaired by filling the damaged area with a latex underlayment compound.

Wood Subfloor

Wood subfloors are constructed in two basic ways, either with plywood sheets or boards. The plywood sheets or boards are nailed to the joists, fastened directly to a concrete surface, or nailed to screeds fastened to a concrete surface.

Make sure all panels or boards are securely fastened to the screeds or joists. If there are loose areas, nail them down with annular-ring nails. Repair any damaged surface areas with wood putty. The wood surface must be clean, dry, and level.

It is very difficult to make wood boards smooth and level enough for resilient floorcovering. Wood board floors should be covered with ¼ or ½-inch exterior grade plywood or underlayment grade hardboard to provide a smooth, level surface. Stagger the panels by nailing them perpendicular to the original subfloor, leaving $\frac{1}{16}$-inch gaps between panels to allow for expansion. Also leave a $\frac{1}{8}$-inch space between panels and walls around the perimeter of the room. Use annular-ring nails to fasten the panels. Nail through the boards into the joists wherever possible.

Existing Floorcoverings

Several types of existing floorcoverings make suitable subfloor surfaces for resilient flooring, provided they are adhered well to the the subfloor, level, clean, and in good repair. If they are not, they must be removed. Remove as much of the floorcovering as possible with a scraping device, taking care not to damage the existing subfloor. Remove all old adhesive with the appropriate solvent. If the existing floorcovering is difficult to remove, a layer of plywood or hardboard can be nailed over the existing surface as described above.

Resilient Flooring If the existing resilient floor is not cushioned and is attached firmly to the subfloor, new resilient flooring can be installed over it. Clean the old floor thoroughly and allow to dry. A coat of primer may improve the bond if adhesive will be used.

Wood Flooring If the existing wood floor is in good condition and level, it can be used as a subfloor for new resilient flooring. The finish must first be removed by sanding if you will be using adhesive to install the resilient floorcovering. It is often easier, but more expensive, to attach a new layer of $\frac{1}{4}$-inch exterior plywood or underlayment grade hardboard over the existing floor.

Ceramic Flooring In most cases, do not install resilient flooring directly over a ceramic tile floor. Ceramic tile should be removed.

Installing Resilient Flooring

As mentioned earlier in this chapter, resilient flooring can be installed either with adhesive or with staples. Resilient floors are easy to install, well within the skill levels of most homeowners. In this chapter, step-by-step installation instructions and photographs are provided for both resilient tile and sheet flooring.

Make sure the subfloor is properly prepared and that all area measurements are accurate before beginning the actual installation.

Tools and Supplies

One of the factors that makes resilient floorcovering installation an easy do-it-yourself project is that the tools required are common household tools which most people have had experience using.

- Common tools - claw hammer, utility knife or linoleum knife, putty knife, heavy-duty scissors or shears, a staple gun and staples (for sheet flooring installation), and a notched trowel (for sheet flooring and resilient tiles installed with adhesive).
- Measuring tools - chalk line and chalk, steel straightedge, carpenter's square, flexible tape measure, and a contour gauge.

If you will be using adhesive, you will need the required amount of adhesive and the solvent recommended by the adhesive manufacturer. Some tile manufacturers recommend the use of a large roller to help set the resilient flooring in the adhesive. If so, rollers can usually be rented from a tool rental company or floor dealer.

Installing Resilient Tiles

Resilient tiles are installed in much the same way as ceramic or wood-parquet tiles. The first installation step is to establish accurate working lines. Do not begin by simply laying tiles at one wall. This can cause the floor to appear off center. Resilient tiles can be laid in a square pattern with tiles running parallel to the walls or in a diagonal pattern with tiles at a 45-degree angle to the walls. The working lines are established as follows:

1. Measure each wall of the area to be floored and mark the midpoint of each wall on the floor. Snap chalk lines from each midpoint to the opposite midpoint, forming two intersecting lines.
2. Check the squareness of the angles. If the lines do not intersect at exactly a 90-degree angle, the room is not square. If they do intersect at a 90-degree angle, everything is fine. The quickest test for squareness of the working lines is to measure 4 feet along one line from the intersection and 3 feet along the other line. If the distance between these two points is exactly 5 feet, the lines are true. If the lines are not true, use a carpenter's square to adjust one of the lines.
3. If a diagonal pattern is desired, mark four points on the working lines, with each point 4 feet from the intersection of the two lines. Tie a pencil to a string. Cut the string 4 feet from the pencil. Place the end of the string on one of the points and draw arcs above and below the point; repeat for all points. The intersecting arcs that will be created indicate diagonal points. Stretch and snap a chalk line between these points. Use the test for squareness described above to determine if the intersecting diagonal lines are square. The diagonal pattern is recommended in corridors and rooms where the length is more than $1\frac{1}{2}$ times the width. Diagonal placement minimizes expansion during high humidity conditions.

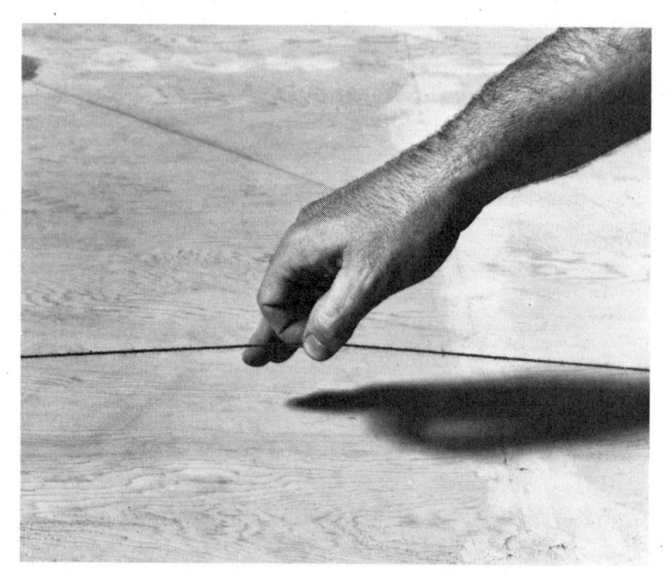

1. Snap a chalk line to indicate working lines.

2. Lay loose tiles and readjust the working lines.

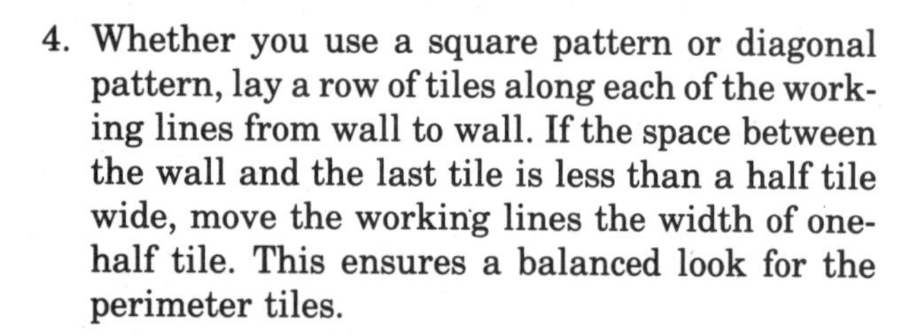

4. Whether you use a square pattern or diagonal pattern, lay a row of tiles along each of the working lines from wall to wall. If the space between the wall and the last tile is less than a half tile wide, move the working lines the width of one-half tile. This ensures a balanced look for the perimeter tiles.

Applying the Adhesive

Assuming that the subfloor is prepared properly, all working lines have been established, and that you have enough tile on hand, it is time to apply adhesive if you are using dryback resilient tile. If you are installing selfstick resilient tile disregard this section and move on to placing the tiles.

1. Carefully read all instructions on the adhesive container. The adhesive should be stored at 70° F for 24 hours prior to use. Check the "open time" for your adhesive. Open time is the length of time during which the tiles can be laid after spreading the adhesive.
2. Spread the adhesive evenly with a notched trowel held at a 60-degree to 80-degree angle. Only cover one quadrant of the working lines at a time, taking care not to spread adhesive over the chalk lines. Apply no more adhesive than can be covered with tiles in 4 hours.
3. If too little adhesive is applied, poor bonding will result. If too much is applied, adhesive will ooze up through the joints between tiles.
4. Allow the adhesive to set for the proper time as recommended by the adhesive manufacturer.

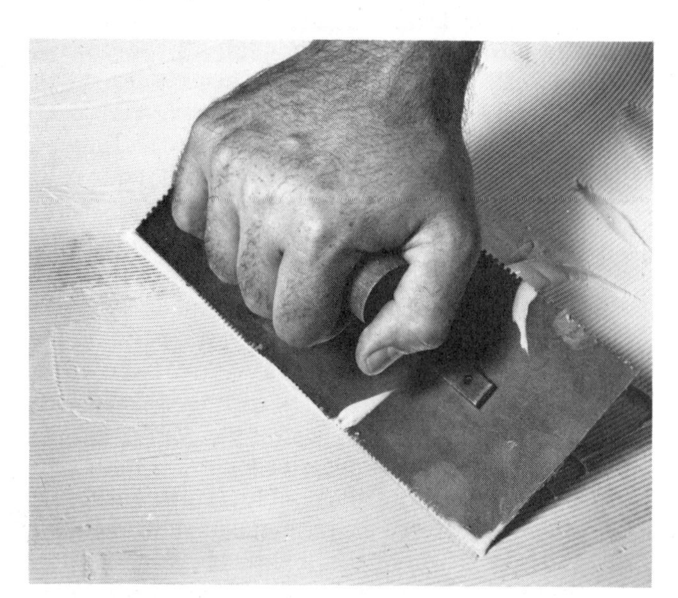

3. Apply adhesive to one quadrant.

4. Lay resilient tiles in a pyramid sequence.

Placing the Tiles

Dryback tiles and selfstick tiles are placed into position in the same way. Like most other tiles, resilient tiles are installed in a pyramid sequence to minimize the chance of misalignment.

1. Place the first resilient tile carefully at the intersection of the working lines in the center of the room. Align edges with working lines; press lightly into place. Never slide the tile. Sliding can cause the adhesive, if used, to collect on the leading edge of the tile which can interfere with a proper fit. The first tile is the key unit. All other tiles are laid off this first tile. Careful, accurate placement is essential for a professional-looking job.
2. Lay the next tiles ahead and to the right of the first tile along the working lines. Continue this stairstep sequence, watching carefully the corner alignment of new units with those already in place. Install one quadrant at a time; leave trimming for later. Many selfstick tiles have arrows on the back side; lay tiles with all arrows facing in the same direction.
3. Remove a tile occasionally to make sure the adhesive, if used, is bonding to the tile properly.
4. Place a sheet of plywood over the laid tile to prevent foot traffic from sliding tiles out of place.
5. Repeat the above procedure for each quadrant until the entire area is covered, with the exception of border tiles.
6. At this point, some manufacturers recommend rolling the floor with a heavy roller to ensure proper mastic adhesion.

6. Resilient tiles can be cut to fit around most obstacles.

7. Use a contour gauge to measure irregular shapes.

5. Measure all border tiles.

8. When all border tiles are cut, apply adhesive and position carefully.

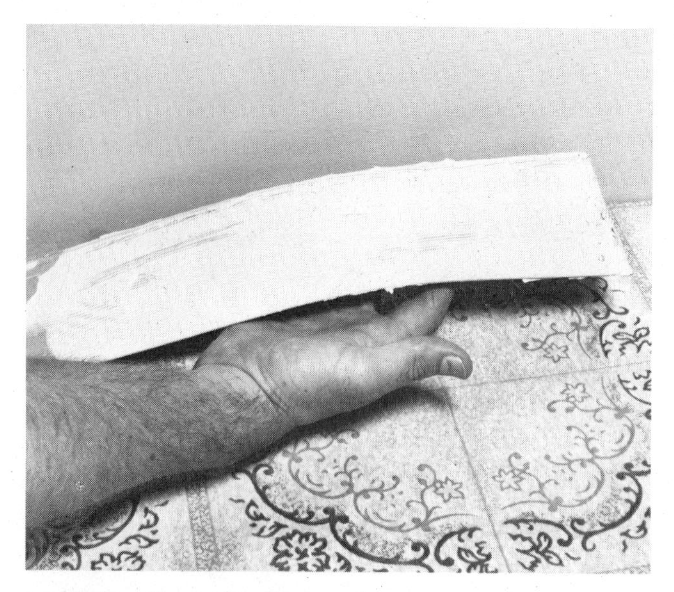

1. *Apply adhesive to the cove base.*

7. At the wall you will have to cut the resilient tiles to fit. Allow ⅛-inch expansion space between the edge of the last tile and the wall. Lay a loose tile on top of the last full tile in the row. Place another tile on top of this tile and slide it over until it almost butts the wall. Mark the first loose tile and cut it with a shears or a utility or linoleum knife. Use a straightedge for straight cuts. Use a contour gauge for curves or other irregular cuts.
8. Allow the flooring to set at least 24 hours before walking on it or placing heavy furniture on it.

Vinyl cove base is a flexible molding that protects walls and gives a finished appearance to a room. It generally comes in 4-foot lengths that butt together with hard-to-detect seams.

Before applying vinyl cove base, be sure all pieces of old molding and any old adhesive are removed and

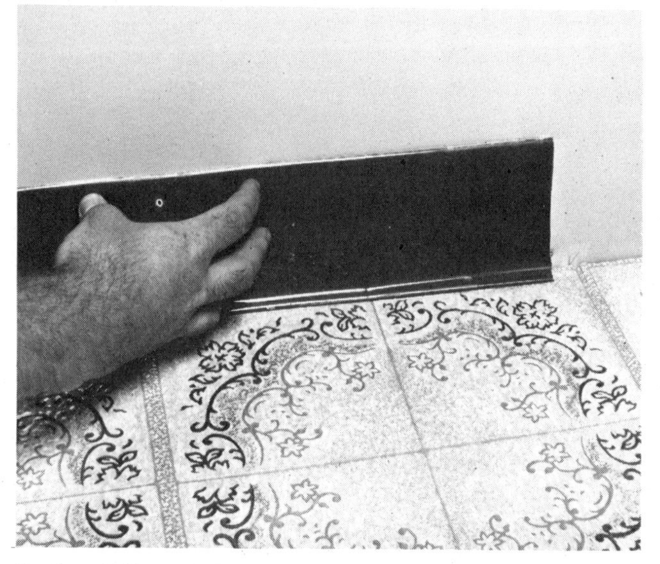

2. *Carefully position the cove base on the wall.*

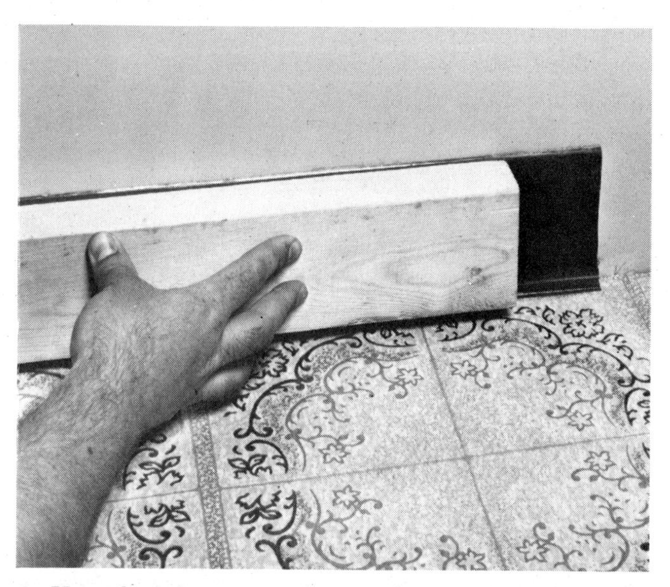

3. *Use a board to secure the cove base to the wall.*

4. *How to fit an inside corner.*

5. *How to fit an outside corner.*

the surface is smooth, dry, and free from oil, grease, loose paint, or other foreign matter. Do not install vinyl molding over wallpaper on walls or paint that has not been dry for at least 2 weeks.

Apply adhesive to the ribbed backside of the cove base with a putty knife. Leave a ¼-inch space at the top edge of the base. Do not spread adhesive on the bottom edge where it curves. Immediately press the cove base to the wall, making certain the bottom edge is on the floor, so you do not have to slide the cove down the wall. The cove base will cover the expansion space between the last row of tile and the wall. Be sure the entire cove surface is pressed firmly against the wall. A wallpaper seam roller or glass jar will help. Then press the toe of the base firmly against the wall, using a straight piece of wood.

1. Cut a floor pattern for the entire area to be floored.

Installing Resilient Sheet Flooring

The key to a professional-looking resilient sheet goods floor is careful planning and accurate trimming. Whether you are using adhesive or staples to secure the floorcovering to the subfloor, the installation procedures are basically the same.

If you haven't already done so, draw an accurate map of the area to be floored on graph paper, indicating all exact measurements. Next figure out the way to lay the flooring with the least amount of seams. Sheet goods generally come in widths of 6 to 12 feet.

In recent years some manufacturers have provided a large piece of pattern paper on which you can draw the room dimensions to exact size. This minimizes the chances of making a mistake when transferring the measurements on the scale drawing to the sheet flooring.

1. Refer to your detailed diagram and roll out the sheet goods flat. Measure and indicate on the sheet flooring lines where all cuts are to be made. Always double-check measurements before making any cuts. Allow 4 inches on each edge for proper trimming.
2. With a utility or linoleum knife or a heavy shears, cut the sheet flooring. Use a straightedge if necessary. Remember to cut all edges 4 inches wider than the actual measurements. If you are cutting over the existing floor or concrete, place one or two thicknesses of cardboard beneath the sheet goods to prevent damaging the floor and dulling the cutting tools.
3. Roll out the cut sheet goods along the longest straight wall. Do not trim until you are certain the material is positioned properly within the area to be floored.

2. Remove the pattern. Transfer it to the sheet floor. Cut.

3. Fold back the floor and apply adhesive.

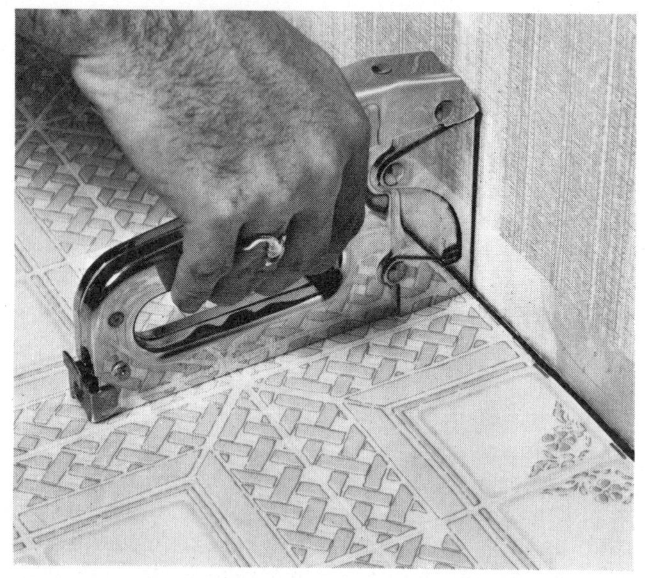

4. Sheet flooring can be stapled to a wood subfloor.

5. Cut all seams carefully.

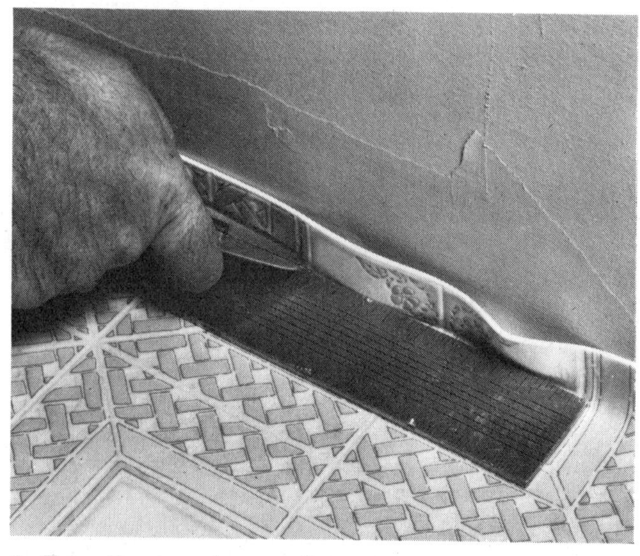

6. Trim flooring along walls.

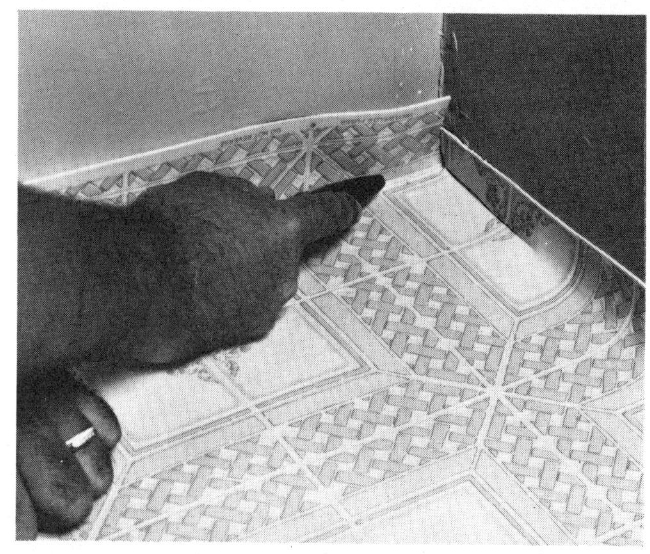

7. How to trim inside corners.

8. How to trim outside corners.

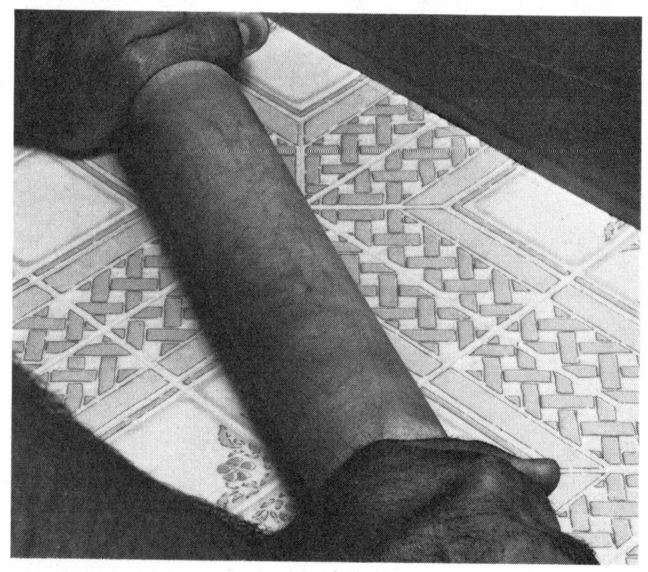

9. Use a roller to set flooring in adhesive.

4. Some sheet vinyl floors require stapling or gluing. If you are using adhesive, roll up half of the flooring to the center of the room. Apply adhesive to the exposed subfloor area as described for resilient tiles. Allow the adhesive to set for the time recommended by the manufacturer, then roll out the floor onto the adhesive and press into place. Roll back the other half and repeat the procedure. Work slowly and carefully.
5. After the entire floor is in place, trim the edges with a utility or linoleum knife. To ensure proper bonding and to remove any air pockets, roll over the surface with a heavy roller.
6. Stapling, if recommended by the sheet goods manufacturer, is much simpler. After Step 3 simply staple the edges of the floorcovering as close to the wall as possible. The cove base trim will hide the staples.
7. If seams are unavoidable, you will have to try to make them as invisible as possible. The easiest way to do this is to overlap the two adjacent pieces of flooring by at least 2 inches. If the flooring has an elaborate pattern or geometric design, make sure the designs on both pieces are perfectly aligned. Using a straightedge and knife, cut along the center of the overlap making sure to cut through both thicknesses of flooring; remove the excess piece. Both sides of the seam should be secured, preferably with adhesive or else with double-faced tape. Lift up the edge and apply adhesive or insert tape. Position edges to create a clean, barely noticeable seam. Clean any adhesive from the surface of the seam with the proper adhesive solvent. Seam sealer can then be applied to fuse the two pieces.
8. Trim all edges, leaving a ⅛-inch gap at the walls. At outside corners, cut flooring along corner. At inside corners cut a vee. With either method, the floorcovering should lie flat at the corner. For a professional-looking job, the flooring must fit tightly around doorways and other obstacles. You can use a handsaw to saw away a thin section of door molding; then slide the flooring into this space.
9. Apply cove base trim as previously described.

Care and Maintenance

Once the floor has been installed, allow 3 or 4 days for the adhesive, if used, to set and establish a good bond. The floor can then be swept or damp mopped. Two problems that often occur are infrequent cleaning and over polishing. Dirt has a very abrasive effect on resilient floors, so sweep or vacuum regularly, every day if necessary. An occasional cleaning with a mild

Resilient tiles can become cracked or damaged with age or heavy use.

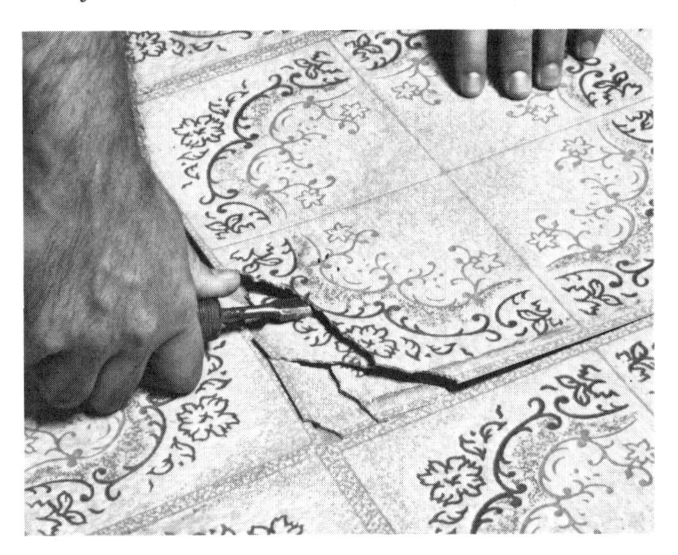

Loosen the damaged tile carefully with a chisel or putty knife. Clean old adhesive from subfloor.

Apply adhesive to the back of a replacement tile. Carefully work the new tile into position.

cleaning solution, such as ammonia and water, and a sponge mop will keep the floor looking like new. Never flood the surface with water.

Today most resilient floorcoverings have no-wax surfaces designed to remain shiny without polishing. Only use products recommended by the manufacturer on these surfaces. If your resilient floor does not have a no-wax surface, you will have to wax it periodically to protect, seal, and add luster to the surface. Too much waxing, however, can cause a dull, yellowish buildup.

Floor-Care Tips

If the floor finish dulls or wears off in the traffic lanes or pivot points of large open floor areas, these specific areas may be cleaned and a light polish applied without redoing the entire floor.

Heavy traffic marks or deep scratches on smooth surface (not embossed) floors may be removed or diminished by burnishing with a synthetic scrub-pad dampened with a mild cleaning solution. Gently rub in even circular motions. Rinse thoroughly and apply recommended floor finish.

Protect resilient floors from direct exposure to sunlight temperatures over 100° F. Protect the floors accordingly with shades, blinds, or drapes drawn over windows or doors during peak sunlight hours.

Oil mops, oil sprays, or sweeping compounds containing oil should not be used on resilient floors.

If it is necessary to move heavy objects, such as furniture or appliances over resilient floorcovering, place panels of plywood or hardboard over the floor to prevent the surface from being marred, cut, or damaged. "Walk" or roll the furniture or appliance across the panel, moving the panel as you proceed. Another method that can be used with lighter objects is to place scraps or pieces of carpet, face-down, under each bearing surface. This will permit the object to slide across the floor.

In areas where grit, soil, and moisture are continuously being tracked in from outside, use a nonrubber or nonlatex backed doormat or rug near the outside doorway.

Always wipe up spills as soon as possible before they dry and stain the floor.

Stain Removal

Obviously the easiest way to eliminate staining your floor is to blot or wipe up spills as soon as possible after they occur. Most spills can be removed with a cleaner and a clean cloth. If certain substances are overlooked or remain on the floor for an extended period of time, however, additional effort may be required.

If the appropriate steps have been taken and the surface appears to be permanently damaged, you may elect to remove the affected tile or section of sheet flooring and replace it.

Follow the procedure steps in the order in which they appear. They may be repeated if you experience difficulty in removing the stain.

Stain Removal Chart

Type of Stain or Spillage	Removal Steps
Tar, asphalt, grease, oil, wax, chewing gum, crayon	A,B,F,G
Ink, lipstick, mustard, catsup, iodine, Mercurochrome, Merthiolate, alcoholic beverages	A,B,C,G
Paint (oil base), varnish, shoe polish	A,B,D,G
Nail polish, lacquer, airplane glue	A,B,E,G
Lye, drain cleaners, detergent concentrates	A,B,G
Rust	B,G

Description of Removal Steps

A. Blot up fresh spills with clean absorbent white cloth or paper towel. Remove dry or hardened substance with stiff putty knife or plastic spatula, taking care not to scratch floor.

B. Wash with clean absorbent white cloth or sponge. For a heavy residue, wet a nylon scrub-pad (recommended for nonstick pans) with cleaner and scrub area. If the stain remains, proceed with next instruction step.

C. Wet clean absorbent white cloth with rubbing alcohol* and rub stain, rotating cloth. Do not walk on area for 30 minutes. If stain still remains, repeat step using liquid chlorine bleach.

D. Wet clean absorbent white cloth with turpentine* and rub stain, rotating cloth. Do not walk on area for 30 minutes.

E. Same procedure as D, except wet cloth with nail-polish remover.*

F. Same procedure as D, except wet cloth with lighter fluid.*

G. Rinse treated area with water; allow to dry. Apply floor finish normally used.

*CAUTION: These solvents are highly flammable. Use appropriate care. Do not smoke or use in vicinity of open flame. Provide adequate ventilation.

Wood Flooring

Wood has long been the traditional floorcovering material for American homes. In fact, many of the beautiful wooden floors that grace historic New England homes built in the the late 1600s and early 1700s are the original finish flooring. Wood was widely used for flooring in America's early days because it was readily available, abundant, and, therefore, inexpensive. But wooden floors certainly were not new. It was the craftsmen from Europe, where wood had been used as a floorcovering for centuries, who brought to the New World the skills and experience necessary for installing wooden plank floors or elegant parquet floors.

Today, wood is no longer an inexpensive floorcovering, yet it continues to be a popular choice for home construction and remodeling projects because of its natural beauty, value, and amazing durability. Wood floors are an excellent investment because they will increase in value over time. This and the permanence of the material offset the moderately high initial installation cost. Installation of a wood floor no longer requires the services of a professional craftsman. Many manufacturers have designed their strip and plank flooring products for easy do-it-yourself installation. Many planks and strips are available prefinished. Wood parquet tiles are as easy to install as vinyl tile. This offsets higher material costs.

Wood is extremely durable. A properly installed and maintained wood floor will last as long as the home or longer, as evidenced by the beautiful floors in colonial homes and European dwellings that are hundreds of years old. Should the surface of a wood floor become distressed or worn, it can be sanded and refinished to look like new. Pound for pound, wood is a very strong material. It withstands shock without damage to the surface.

Unlike another extremely durable floorcovering, ceramic tile, wood offers good resiliency. Wood fibers have a natural elasticity that makes walking on the surface comfortable. Wood is made of thousands of tiny hollow pores that trap air, giving wood excellent insulating qualities. It does not feel cold underfoot because it does not absorb body heat.

Much of the allure of a wood floor is due to its inherent beauty. Wood has a natural warmth and richness that is impossible to duplicate with synthetic materials. This unique enduring beauty can only be created by nature over a period of many years. Wood not only looks beautiful, but it feels beautiful as well. The subtle textures of the grain can be felt by simply passing a hand over the surface lightly. Every piece of wood is in itself unique. This truly appeals to man's aesthetic sense.

Whereas most floorcoverings become less beautiful with age and wear, wood can actually become more beautiful. With time, wood develops a rich, mellow golden tone called patina. This impressive quality is a gradual darkening and aging of the wood that cannot be duplicated by staining or refinishing. The color change is due to exposure to light and air. This warm, honest beauty never goes out of style and is appropriate to most interior designs and styles, from rustic early American to elegant contemporary.

The Choices

When selecting wood floorcovering materials, there are more factors to consider than one might realize. Several species of both hardwoods and softwoods are suitable for floors. The terms hardwood and softwood are general names and do not have a direct bearing on the wood's actual strength and hardness.

Certain hardwoods are more commonly used for flooring because they tend to have higher resistance to wear, more uniform wearability, are less likely to sliver, and have better grain characteristics making them more acceptable for finishing materials. The two most common flooring hardwoods used in residential construction are oak and maple. Other hardwoods used in flooring include: beech, birch, ash, hickory, teak, pecan, cherry, walnut, and several other species. Exotic woods such as teak, mahogany, walnut and cherry are often used in parquet floors.

There are twenty species of oak trees used for flooring, grouped as either white or red oak. There is very little actual color difference, contrary to what the names may indicate, and the quality is virtually the same for both groups. There is primarily one species of maple most commonly used for flooring—sugar maple. Maple is smooth with very little grain pattern or texture. Oak has much greater character with dramatic grain patterns that give it beauty. Where a smooth uniform surface is required, however, maple may be the superior floorcovering wood.

Softwoods most commonly used as flooring materials are southern or yellow pine, Douglas fir, western larch, and western hemlock. Woods such as redwood, cedar, southern cypress, and eastern white pine are commonly used in areas where the particular species is abundant. Softwoods are usually less expensive than most hardwoods because they are not as strong or durable; however, they are perfectly suitable for flooring in low traffic areas such as bedrooms and closets.

Every wood, whether hardwood or softwood, is an organic material that has its own set of identifying characteristics that distinguish it from other woods. The most important characteristics are color, hardness, strength, grade, and grain. Both appearance and cost depend on two other factors: how the wood was sawed from the log and in what form it is manufactured, such as parquet, plank or strip, unfinished, or prefinished. Listed below are the various factors to consider when selecting wood floorcoverings.

Color Each wood has its own natural color ranging from creamy white to chocolate brown; however, the actual color of the finished floor is usually not the natural color of the wood. Through finishing the natural color can be enhanced, strengthened, bleached, or even completely changed. Prefinished wood flooring has been stained, sealed, and finished. Unfinished flooring can be stained any color or tone. The degree of color is determined by the use of stains, bleaches, pigmented fillers, or pigmented finishes. Even the clear top protective coating can alter color somewhat. Finishes are available in several glosses ranging from a soft satiny appearance to a very high sheen. Most woods can take on a wide variety of beautiful tones. Select a color that complements the style of your home or furnishings.

Strength and Hardness The ability of wood to withstand shock without damage is called hardness. Strength is the amount of weight the wood can consistently support. Most hardwoods are very strong, hard, and durable. Softwoods are only slightly less strong and hard.

Grain Good grain characteristics are a very important consideration because it is grain that gives wood character and beauty. There is no substitute for good grain pattern. Stain can change color, veneers can improve hardness, and the surrounding framework can increase strength; but for wood to be beautiful, the inherent grain pattern must be beautiful. The appearance of grain through the finish enhances beauty. Grain is formed by pores in the wood. The pattern these pores form when wood is cut is called figure. Figure is largely dependent on the sawing method used to create the boards. Sawing method can affect cost as well. The most common sawing method is plain sawn. The figure of a quarter sawn log is more prominent, but quarter sawing takes longer and creates more waste, so it is more expensive. Wood is open-grained if the pores are visibly open. Close-grained wood has tightly packed pores not visible to the naked eye.

Grade Grade affects cost. Appearance alone determines the grade of wood. Wood that is generally free of defects or imperfections is called "clear". This grade may contain burls, streaks, and pinworm holes. "Select" wood contains more natural imperfections such as knots. "Common" grades have more markings and variations in color.

Manufacturing As mentioned earlier in the book, wood flooring is available as strips, planks, parquet wood tiles, or blocks, depending on how it is milled. Each of these types is available prefinished or unfinished. All wood floors must be finished with a protective coating. Applying the finish yourself may save money. Wood flooring is usually tongue and grooved although square edge is available. Wood parquet tiles are easy to install. Plank and strip floors are difficult for the average homeowner to install although manufacturers are introducing products designed for do-it-yourself installation. Wood flooring can be nailed into place, glued with adhesive, or held in place with selfstick backs.

Forms of Wood Flooring

Strip flooring is the basic type of wood flooring used in residential construction. Strip flooring consists of narrow boards of random lengths, 2 to 8 feet long, locked together with tongue-and-groove edges. Some strip flooring boards are available with square edges. The boards vary in width from 1½ to 3½ inches with 2¼ inches being the most common width. Thickness varies from 5/16 to 7/8 inch, with 3/4 being the most common. Many tongue-and-groove boards have a tongue on one end and a groove on the opposite end. This is referred to as end-matched and is an excellent feature if you plan to install strip flooring yourself. Many strip flooring boards have relatively wide shallow channels cut on the back face to add greater resiliency and facilitate installation over subfloor imperfections.

Plank flooring is very similar to strip flooring except that the planks are generally wider, usually between 3½ and 8 inches. The planks often vary in both length and width. Plank flooring is the oldest flooring material used in America. Not unlike strip flooring, plank flooring is available with or without tongue-and-groove edges and end-matching, and it can be installed in the same way. Plank flooring, however, can be laid in adhesive as well. Wider planks should be installed over a plywood subfloor. The traditional method of fastening the planks to the subfloor, dating back to colonial times, uses wooden pegs. Today, many styles of plank floors have simulated or real wooden plugs.

Parquet flooring, commonly called block flooring, is a wood flooring laid in blocks or squares. This type of wood flooring has experienced an amazing growth in popularity over recent years because it is price competitive with other types of floorcoverings, is commonly available in a wide variety of woods, finishes, styles, and patterns, and is easy for the novice to install. Parquet flooring is actually solid wood tiles, sheets laminated with hardwood veneer, or small inlaid pieces of wood that create a pattern or mosaic. Small pieces of wood are generally bonded to a backing of wood or mesh. In fact, some manufacturers actually call parquet "wood tiles" because they are laid in adhesive much like resilient floor tiles. Most parquet tiles are 6 or 12-inch squares ranging in thickness from 3/8 to 7/8 inches. Some manufacturers make parquet flooring with tongue-and-groove edges and/or with an adhesive backing. Parquet is generally made from hardwoods such as oak, ash, maple, walnut, pecan, and teak, and is available finished or unfinished. Actual color is achieved through finishing. Factory finishes are usually baked into the wood to create a durable, easy-to-maintain surface. Although less expensive and available in more patterns than prefinished tiles, unfinished parquet tiles require tedious time-consuming work to properly finish them. Most unfinished parquet floors, as well as unfinished strip or plank floors, must be sanded, sealed, filled, stained, and finished. You then have complete control over color.

Parquet tiles create a rich, elegant-looking floor. Installing wood parquet is well within the skills of most homeowners.

Where to Use Wood Flooring

Wood adds warmth and elegance to any room. Whether you plan to use rustic-looking random-width planks with walnut peg accents, linear strips that create a strong parallel look, or elegant, time-honored parquet tiles that are available in a wide assortment of patterns, tones, and surface textures, wood flooring adds exceptional beauty to any room. Wood provides a suitable setting for furnishing styles from early American to contemporary.

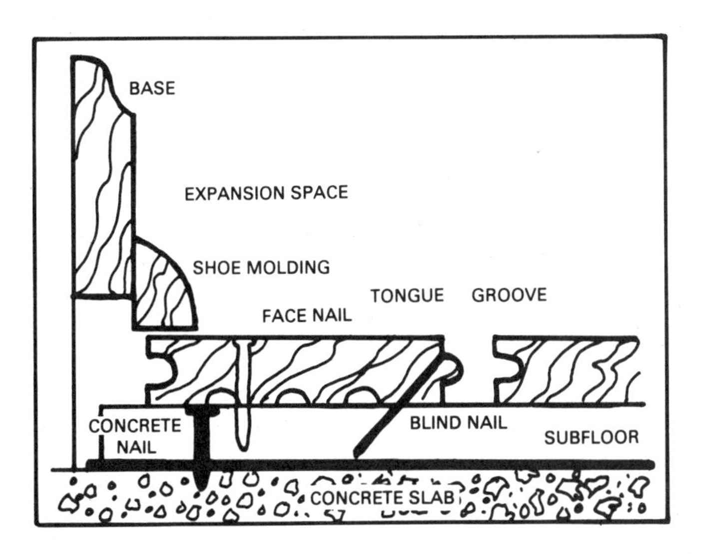

You no longer have to own a house with existing wood floors to enjoy the charm and natural beauty of wood. Many plank and parquet products can be installed over many types of existing surfaces. This makes wood flooring possible in any room of the house.

The primary enemy of wood is moisture. Wood flooring should not be exposed to prolonged periods of moisture or high humidity as wood can warp or rot if subjected to these conditions. It therefore is not too practical for use in bathrooms. Manufacturers in recent years have developed moistureproof finishes and sealers to prevent possible water damage. These products can be used in the bathroom; but be certain of the finish capabilities before applying such a product.

Halls, foyers, and bedrooms have traditionally been popular areas for wood floorcoverings. With the variety of beautiful easy-to-install parquet patterns available today, wood floors are now being used more in kitchens, family rooms, dining rooms, and living rooms. Wood floors enhance furniture and other decorating elements; they blend well with the rich textures of carpeting and wallcoverings. The endless variety of wood tones allows wood flooring to look completely natural with any interior design.

Preparing the Subfloor

Whether you plan to install strip, plank, or parquet wood floorcovering, you must first prepare a suitable subfloor. Several materials make good subfloors for wood floorcoverings, including a concrete slab, a wood subfloor, or some existing floorcoverings. The subfloor is prepared in basically the same way for all three types of wood flooring. Many wood floor problems such as squeaks, cracks, and loose boards can be traced directly to improper subfloor preparation.

For wood floors, the subfloor must do more than provide a sound sturdy base; it must also protect the wood. Water and moisture can severely damage a wood floor in a very short time. Precautions to prevent this must be taken when preparing the subfloor. Any wood expands across the grain when exposed to moisture and contracts during dry periods. Some seasonal movement is to be expected, but try to keep moisture at a consistent level or buckling may result. Water and moisture problems can develop from several areas.

- Water in the basement or under your home can be caused by a poorly constructed drainage system, gutters and downspouts that do not work effectively during rain storms, cracks in the foundation, or leaky and/or sweating water pipes.
- The moisture content of the concrete may be too high.
- Windows and doors that do not fit properly and, therefore, allow moisture to enter the area.
- Wood, whether in the joists, subfloor backing, or the actual floorcovering, that has not been properly cured.
- Excessive atmospheric humidity in the area.

Many of these problems can be avoided or solved. Vapor barriers can be used to protect the floor from excessive moisture. Ventilation can be added to help reduce humidity. Ventilation openings should equal 1½ percent of the first floor area. Repairs can be made to a faulty foundation, leaky pipes, ill-fitting doors or windows, or a poor rain-carrying system. Use only dry lumber for any flooring project. A dehumidifier can help remove moisture from the atmosphere.

Wood flooring also needs protection against excessive heat. Wood can dry out and crack when exposed to excessive heat for prolonged periods of time. If the wood floor is to be laid over a heating plant, uninsulated heating ducts, or radiant-heated concrete subfloor, nonflammable insulation must be used to prevent the wood from drying out.

Preparing a Concrete Subfloor

Concrete makes a suitable subfloor for wood floorcoverings provided it is thoroughly dry, level, and clean. The concrete slab must be constructed properly. New concrete is heavy with moisture. Even older concrete can have a high moisture content, particularly below-grade or on-grade slabs where hydrostatic pressure is present. The first step in preparing a concrete subfloor is to determine the present moisture content of the concrete. This can be done using one of the following tests. It is best to test during spring or other wet periods. Conduct these tests in several areas of the room(s) to be floored.

1. Tape a one-foot square of clear heavy polyethylene film to the slab, sealing all edges with plastic packaging tape or duct tape. Leave this in place for 48 to 72 hours. If there is no clouding or beads of moisture present, the concrete slab can be considered dry enough to install wood floors. If moisture is present, give the slab more time to dry and repeat the test. Drying can be accelerated with heat and ventilation. If moisture is still present, carefully assess the home's drainage and rain-carrying system. Repair any problems. If moisture is still present, select another type of flooring less affected by moisture.
2. Place a quarter teaspoon of dry calcium chloride crystals (available at drugstores) inside a 3-inch putty ring on the slab. Seal the crystals off from the air by covering them with a glass. If the crystals dissolve within 12 hours, there is too much moisture present to install wood flooring.
3. Put several drops of 3 percent phenolphthalein solution in grain alcohol at various spots around the area. If a red color appears in a few minutes, there is moisture present and wood flooring should not be installed.

If the slab is sufficiently dry, make sure the surface is level and clean. Remove all wax, dirt, and grease. Do not use water; use special chemical cleaners for grease and oil. Chip off all raised imperfections or irregularities and fill all depressions, cracks, and holes with a suitable patching compound. Apply a coat of asphalt sealer to the surface for added moisture protection. Concrete sealer or curing compounds must be removed. If the concrete floor is rough and uneven, a relatively thin new slab can be poured over the existing one.

Some types of parquet wood tiles can be applied directly to a properly prepared concrete subfloor that is on or above-grade. Most parquet manufacturers recommend not laying wood tiles on below-grade surfaces or over radiant-heated subfloors. Proper on-grade concrete subfloor construction has a vapor barrier between the ground and concrete slab to prevent the evaporation of moisture through the slab. For most types of wood flooring, however, it is recommended that the floorcovering be installed over wood secured to the concrete subfloor.

Preparing a Base over a Concrete Slab

Wood flooring, particularly strip and plank flooring, requires a wood nailing base for successful installation over a concrete slab. A good base can be constructed in one of two basic ways, either using the plywood-on-slab method or the screeds method. Both methods require a dry, level, and clean concrete surface.

Plywood-on-Slab Method This method provides a suitable nailing base for strip, plank, and parquet flooring that cannot be installed directly over a concrete subfloor. Check with the manufacturer's directions if there is any question.

1. Seal the clean dry slab with asphalt sealer if you have not already done so.
2. Cover the sealer with a layer of 4 mil or heavier polyethylene film, overlapping edges 4 to 6 inches and allowing enough to extend underneath the baseboard on all sides. This serves as a vapor barrier to retard evaporation. The edges do not have to be sealed.
3. Lay plywood panels out loose over the entire floor. Use ¾-inch exterior grade plywood. Cut the first sheet of every run so end joints will be staggered 4 feet. Leave ½-inch space at all wall lines and ¼ to ½-inch between panels. At doors and other vertical obstructions where molding will not be used to cover the void, cut the plywood to fit, leaving about ⅛-inch space.
4. Fasten the plywood with a power-actuated concrete nailer or hammer-driven concrete nails. Use a minimum of nine nails per panel, starting at the center of the panel and working toward the edges, to be sure of flattening out the plywood and holding it securely.

This will provide a solid subfloor which can now be treated like any other wood subfloor for purposes of installing wood floorcovering.

Screeds Method This method uses screeds (also called sleepers or nailers), suitable as a nailing base for strip and plank flooring to widths of 4 inches. Screeds are short sections of No. 1 or No. 2 common 2×4-inch boards. The 2×4-inch boards should be pressure-treated with chemicals for protection against moisture and wood-boring insects. Do not use creosote because it can stain the wood.

1. Cut the treated 2×4-inch lumber into random lengths from 18 to 48 inches. The screeds must be flat and not warped.
2. Seal the dry, clean slab with asphalt sealer if you have not already done so.
3. The screeds should be installed at right angles to the direction you plan to install the floor.
4. Apply hot asphalt mastic to the sealed surface and inbed the screeds, 12 inches on center. Stagger joints, overlapping ends 4 to 6 inches. Leave ¾-inch space between ends of screeds and walls.

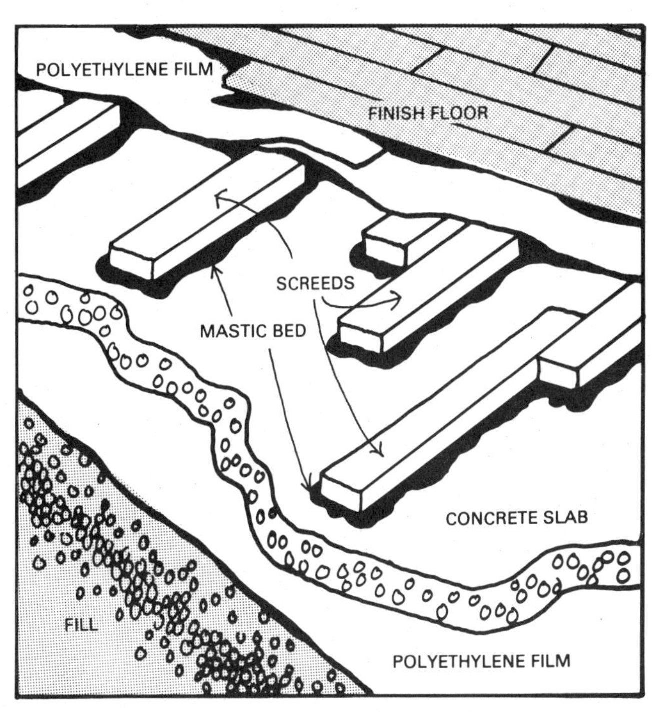

One way of installing a wood floor on a concrete slab is by laying wooden screeds in a mastic bed and fastening a wooden subfloor to the screeds. The finish floor can then be installed over the subfloor.

5. When the screeds are set, spread a vapor barrier of 4 or 6 mil polyethylene film over them, overlapping edges at least 6 inches. Avoid bunching and punctures. Sealing the edges is not necessary. Finish flooring is nailed through the film into the screeds. An alternative vapor barrier can be created by applying two layers of 15-pound asphalt paper to the primed concrete slab with mastic and then installing the screeds over the asphalt paper. A polyethylene layer over the screeds is still recommended.
6. Nail strip or plank flooring to the screeds.

If the strip or plank flooring you intend to use is 4 inches wide or wider, the subfloor should be ⅝-inch plywood-on-a-slab or ¾-inch boards,6 inches wide,installed over the screeds. The finish flooring is then nailed to the plywood or board subfloor. This provides an adequate nailing surface and a more structurally sound subfloor.

Depending on the manufacturer's recommendations, some parquet wood tiles can be applied directly to a concrete subfloor with mastic; however a vapor barrier between the subfloor and finish flooring is required. If your parquet tiles can be laid using this method, follow these steps:

1. Prime the concrete surface with asphalt sealer and allow to dry, if you have not already done so.

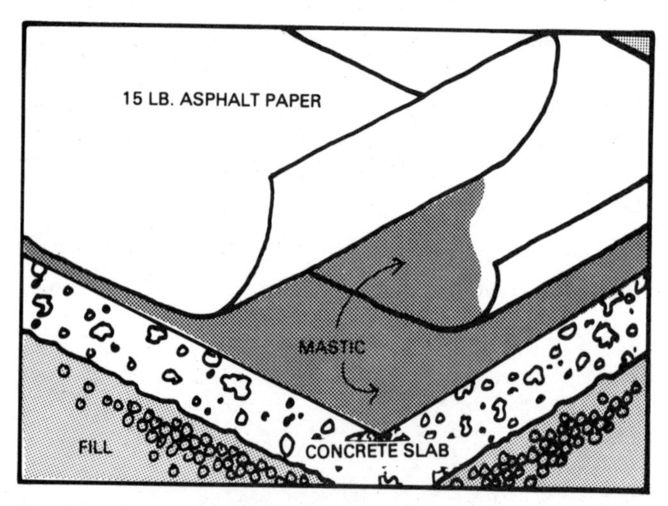

Double layer of asphalt paper moisture barrier.

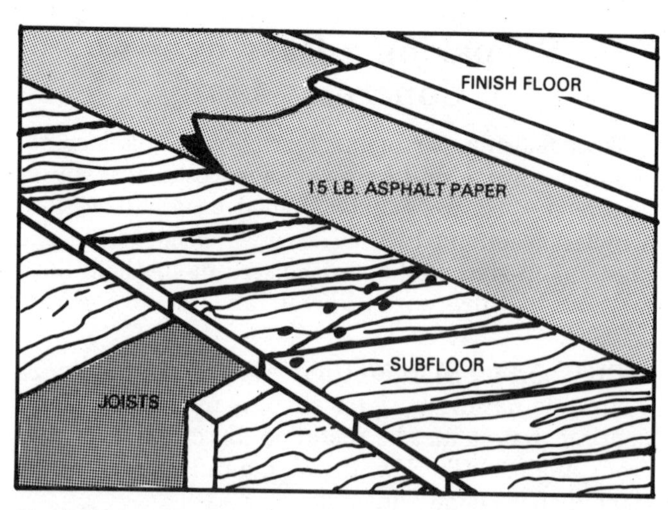

Single layer of asphalt paper moisture barrier.

2. Apply cut-back asphalt mastic to the entire slab surface with a straightedge trowel and allow to dry 30 minutes.
3. Unroll 4 or 6 mil polyethylene film, overlapping edges 4 inches. Cover the entire surface.
4. Walk over the entire surface to inbed the film and ensure proper adhesion. Small bubbles are of no concern.

Another method is the two-membrane asphalt paper method. This produces an excellent vapor barrier.

1. Apply mastic to primed concrete slab surface with a notched trowel and allow to set 2 hours.
2. Roll out 15-pound asphalt paper, overlapping edges 4 inches and butting all ends.
3. Apply more mastic in same manner.
4. Apply another layer of asphalt paper in same manner. Stagger the overlaps.

Finish flooring is then applied directly to the vapor barrier with mastic.

Preparing a Wood Subfloor

Wood subfloors are generally either constructed of ½ or ¾-inch exterior grade plywood or 1×4 or 1×6-inch No. 1 or No. 2 common square-edged lumber applied over wood joists. Again, the subfloor surface must be dry, level, clean, and structurally sound. Repair any wooden subfloor problems. This usually requires removing any existing floorcovering. Proper nailing keeps boards rigid and prevents squeaking. Make sure all boards or plywood panels are secured well to the floor joists. Renail any loose lumber and replace severely damaged lumber. If the subfloor is in poor condition or is constructed of narrow boards, install a layer of underlayment. This is generally ¼-inch plywood applied at right angles to the existing subfloor. Nail every 6 inches into a joist, leaving ¼-inch between panels. If mastic will be used, prime the wood subfloor surface to assure proper bonding between adhesive and subfloor. If the new floor will be nailed to the subfloor, cover the subfloor surface with 15-pound asphalt paper first.

Preparing an Existing Floor

Wood floorcovering can be applied directly over some existing floorcoverings. As with any subfloor surface, the existing floor must be level, clean, structurally sound, and in good condition. Begin preparations by removing the shoe moldings and baseboards, taking care not to damage them. Number all pieces so they can be returned to their original positions after installing the new floor.

Existing Wood Floors New wood flooring can be applied over an existing wood floor if it is sound. If you will be nailing wood tile or strip or plank flooring to the floor, countersink any protruding nailheads, nail any loose boards to the joists, and roughsand any high spots or irregular areas. Existing tongue-and-groove flooring should be covered with a layer of ¼-inch plywood underlayment for new parquet installation. Do not apply parquet flooring over existing strip or plank flooring. If the new wood floor will be applied with mastic or adhesive, any finish, stain, wax, or sealer will have to be removed from the existing surface. This can involve considerable effort and time. It is often easier to apply a layer of ¼-inch plywood underlayment over the existing wood floor.

Existing Resilient Floors Wood flooring can be installed over an existing resilient tile or sheet floor, provided the surface is dry, clean, level, and securely bonded to the subfloor. Wood flooring should not be installed directly over cushioned vinyl flooring, rubber tile, or no-wax resilient flooring. If the existing resilient floor is not in good condition, it will have to be removed and the subfloor thoroughly cleaned. If the existing floor is in good condition, roughsand to remove all wax. Roughen the surface to ensure better mastic adhesion if the new wood floor is to be set in adhesive.

Installing Wood Floors

Installation procedures vary for different types of wood flooring. Parquet floors are easy to install, well within the skills of most do-it-yourselfers. Strip and plank flooring on the other hand is very difficult for the do-it-yourselfer to install and often requires special tools and equipment. Depending on your skills and experience, either select a type of flooring that can be installed within these limitations or consider hiring a professional contractor to install the floor for you.

Materials

The first thing you need for installing a wood floor are the proper materials. Before purchasing these materials, make a scale drawing of the area to be floored as described in the chapter on estimating area. This drawing, with all measurements indicated, will help your dealer determine the exact amount of materials needed for the job. Order enough extra flooring materials to allow for waste and provide some extra materials for future replacement should the need arise. Usually 3 to 5 percent extra is plenty.

Unlike other types of floorcoverings, wood needs time to adjust to the existing conditions of the area to be floored before it can be installed. Arrange with your dealer to have the flooring delivered or picked up at least 4 days before installation. Wood flooring should not be delivered in snowy, rainy, or extremely damp or humid weather because the wood will absorb moisture. After the wood flooring is installed in a dry room, it will eventually shrink, causing it to crack or creating gaps between boards.

When your shipment of wood flooring arrives, store it in or near the room in which it will be installed. The storage area must be warm and completely dry, near the levels that will be present after the flooring is installed. Uncarton or unpack the wood flooring and stack boards or tiles loosely to permit air to circulate around them. Some manufacturers seal wood flooring in packages under ideal conditions. This type of product can be installed right after it is unwrapped.

At the time you order flooring, also order all other materials: trim if desired or necessary, adhesive if necessary, adhesive solvent, nails, primer, or any special equipment required for installation. It is best to buy everything from one dealer to ensure compatibility of materials.

Tools and Other Supplies

If you plan to install parquet wood flooring, you will need the following tools. Parquet floor installation does not require very many special tools. Most of the tools and supplies are common household items.

Wood flooring must be stacked loosely in the room in which it will be installed to allow the wood to adjust to the climatic conditions of the surrounding area.

- Common tools—a claw hammer, crosscut saw, coping saw or electric saber saw, and putty knife.
- Measuring tools—carpenter's square, steel tape measure, chalk and chalk line, and contour gauge.
- Special tools—a proper notched trowel to spread the adhesive.

If you have chosen unfinished parquet flooring, you will need all of the materials required to sand, seal, stain, and finish the floor after it is installed. The same applies to unfinished strip or plank flooring.

Since a wood strip or plank floor is more difficult to install than parquet flooring, more tools are required. You can get by using only common household tools, but a couple of specialized tools will make the job significantly easier. The tools needed include:

- Common tools—a claw hammer, crowbar, electric drill and drill bits, nail set, crosscut saw, backsaw and miter box, and rubber mallet.
- Measuring tools—carpenter's square, steel tape measure, chalk and chalk line, and contour gauge.
- Special tools—a power nailer (available at most tool rental businesses or wood flooring dealers). The power nailer drives special nails at the proper depth and angle.

Many planks are fastened to the subfloor with screws. Check the manufacturer's installation instructions. If this is the case, you will need a screwdriver (preferably a power screwdriver) and an ample supply of screws. Assemble all tools prior to installation.

Installing Wood Parquet Floorcovering

The first step in installing a wood parquet tile floor properly is to establish accurate working lines. It is from these lines that the entire installation is based. If they are not accurate, the entire installation may fail. Never begin laying tiles at a wall because most rooms are not perfectly square. Parquet wood tiles are like ceramic or resilient tiles in this respect: working lines are established for a wood parquet floor in exactly the same way as for ceramic or resilient tiles. For more detailed instructions on establishing working lines, please refer to the chapter on ceramic tile installation.

Establish Working Lines

Parquet can be laid in a square pattern with tiles running parallel to the walls or in a diagonal pattern with tiles at a 45-degree angle to the walls. The working lines are established as follows:

1. Measure each wall of the area to be floored and mark the midpoint of each wall on the floor. Snap chalk lines from each midpoint to the opposite midpoint to form two intersecting lines.
2. Check the squareness of the lines. If the lines do not intersect at exactly a 90-degree angle, the room is not square. The quickest test for squareness of the working lines is to measure 4 feet along one line from the intersection and 3 feet along the other line. If the distance between these two points is exactly 5 feet, the lines are true. If the lines are not true, use a carpenter's square to adjust one of the lines.
3. If a diagonal pattern is desired, mark 4 points on the working lines, with each point 4 feet from the intersection of the two lines. Tie a pencil to a string; cut the string 4 feet from the pencil. Place the end of the string on one of the points and draw arcs above and below the point; repeat for all points. The intersecting arcs indicate diagonal points. Stretch and snap a chalk line between these points. Use the test for squareness described above to determine if the intersecting diagonal lines are square. The diagonal pattern is recommended in corridors and rooms where the length is more than 1½ times the width. Diagonal placement minimizes expansion during high humidity conditions.
4. Whether you use a square pattern or diagonal pattern, lay a row of tile along each of the working lines from wall to wall. If the space between the wall and the last tile is less than a half tile wide move the working lines the width of one half tile. This ensures a balanced look for perimeter border tiles.

Most existing parquet patterns can be laid out with these two working lines. Herringbone patterns will require two test lines: one will be the 90-degree line already described and the other crosses the same intersections of lines, but at a 45-degree angle to both.

If such preliminary layout preparation seems a bit elaborate, keep in mind that it is wood you are installing. Each piece must be carefully aligned with all neighboring pieces. Small variations in size, natural to wood, must be accommodated during installation to keep the overall pattern squared. You cannot correct a "creeping" pattern after it develops; a carefully laid-out floor causes less problems during fieldwork.

Applying the Adhesive

Before applying adhesive, make sure you have enough tile to complete the project. Read the manu-

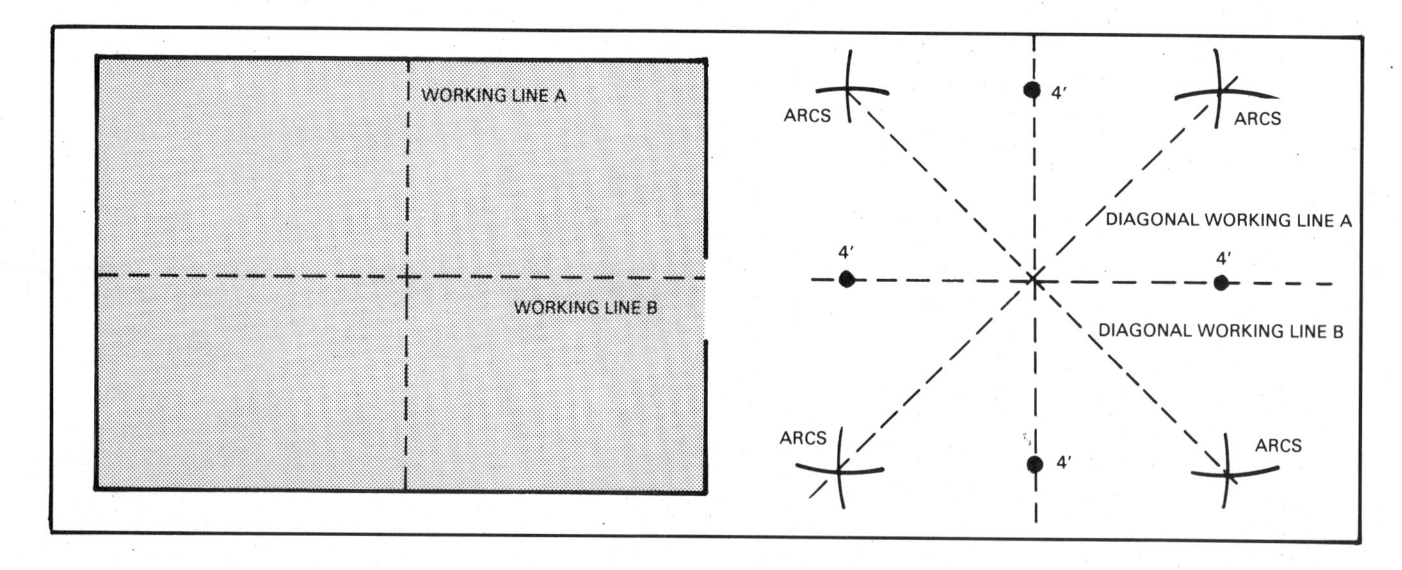

1. Mark all working lines, parallel or diagonal pattern.

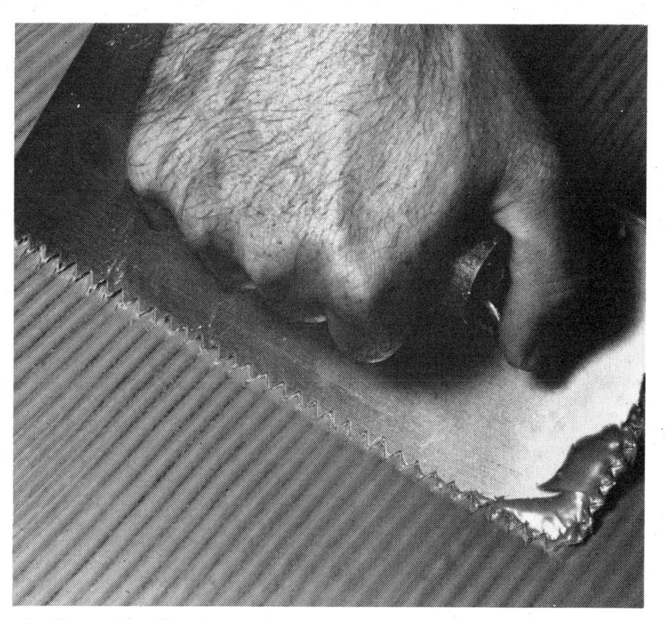

2. Spread adhesive.

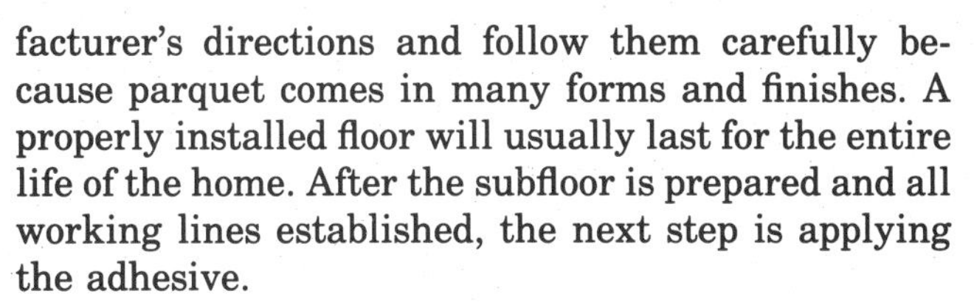

facturer's directions and follow them carefully because parquet comes in many forms and finishes. A properly installed floor will usually last for the entire life of the home. After the subfloor is prepared and all working lines established, the next step is applying the adhesive.

1. Read all instructions on the adhesive container. Most parquet flooring adhesives are spread at a rate of 40 square feet per gallon. The adhesive should be stored at 70° F for 24 hours prior to use. Check the "open time" for your adhesive.
2. Spread the adhesive evenly with a notched trowel held at a 60 to 80-degree angle. Only cover one quadrant of the working lines at a time, taking care not to spread adhesive over the lines. Apply no more adhesive than can be covered with tiles in 20 minutes.
3. If too little adhesive is applied, poor bonding will result. If too much is applied, adhesive will ooze up through the joints between tiles.
4. Allow the adhesive to set for the proper time as recommended by the adhesive manufacturer.

3. Carefully slip tongue into groove.

Placing the Tiles

Wood parquet must always be installed in a pyramid or stairstep sequence rather than in rows. This again prevents the small inaccuracies of size in all wood from magnifying or "creeping" to gain an appearance of general misalignment.

1. Place the first parquet unit carefully at the intersection of the working lines. Align edges with working lines. Press lightly into place. Never slide the tile. Sliding causes the adhesive

4. Align the edges of the tiles carefully.

5. *Install tiles in pyramid sequence.*

to collect on the leading edge of the tile which can interfere with a proper fit. The first tile is the key unit. All other tiles are laid off of this first tile.

2. Lay the next tiles ahead and to the right of the first tile along the working lines. Continue this stairstep sequence, watching carefully the corner alignment of new units with those already in place. Install one quadrant at a time. Leave trimming for later.
3. Remove a tile occasionally to make sure the adhesive is bonding to the parquet properly.
4. Place a sheet of plywood over the laid tile to prevent foot traffic from sliding tiles out of place.

6. *As you work, place a sheet of plywood over the laid tile.*

7. *Roll the floor to set tiles in adhesive.*

8. *Mark all border tiles for cutting.*

9. *Cut tiles with handsaw.*

10. Reducer strips are used on open floor edges.

5. Repeat the above procedure for each quadrant until the entire area is covered, with the exception of border tiles.
6. At this point some manufacturers recommend rolling the floor with a heavy roller to ensure proper mastic adhesion.
7. At the wall you will have to cut the parquet tiles to fit. Allow ½ to ¾-inch expansion space between the edge of the last tile and the wall. Mark the tile and cut it with a crosscut or saber saw. Cork blocking is available for these spaces to permit the flooring to expand and contract.
8. At each doorway, cut off the casing with a crosscut saw, using a tile as your guide. Slide a tile beneath the casing to fit around the door. Allow some expansion space.
9. End the floor beneath the door. If there is an appreciable difference in the height of the two adjacent floors, a reducer strip should be installed.
10. Nail moldings into walls, not into tiles.
11. Allow the floor to set at least 24 hours before walking on it or placing heavy furniture on it.

Installing Hardwood Strip or Plank Flooring

As mentioned previously in this chapter, the installation of strip or plank flooring is considerably more difficult and time-consuming than installing parquet tiles. It requires skill and experience with the tools and materials needed to install this type of floor efficiently and successfully. Provided the subfloor is in proper condition, you can begin installing the plank or strip floor. Tongue-and-groove strip and tongue-and-groove plank flooring are installed in a very similar manner. Square-edge planks are installed with screws, and square-edge strips are installed with nails. Try installing the boards perpendicular to the joists, blind nailing them through the subfloor and into the joists. This technique results in structural strength.

Tips for Easier and Better Wood Flooring Installation

- Work from left to right. In laying strip flooring, you will find it easier to work from your left to your right. Left is determined by having your back to the wall where the starting course is laid. When necessary to cut a strip to fit to the right wall, use a strip long enough so that the cut-off piece is 8 inches or longer and start the next course on the left wall with this piece.
- Save short pieces for closets. For best appearances, always use long flooring strips at entrances and doorways. Save some of the shorts for closet areas and scatter the rest evenly in the general floor area.
- Put a "frame" around obstructions. You can give a much more professional and finished look to a strip flooring installation if you "frame" hearths and other obstructions, using mitered joints at the corners.
- Reversing the direction of strip flooring. Sometimes it is necessary to reverse the direction of the flooring to extend it into a closet or hallway. To do this, join groove edge to groove edge, using a slip tongue available from flooring distributors. Nail in the conventional manner.
- Use only sound, straight boards for subfloors. The quality of the subflooring will affect the finish flooring. Use only square-edge ¾-inch dressed boards no wider than 6 inches. Boards which have been used for concrete-form work are often warped and damp and should not be used.

To install tongue-and-groove strip or plank flooring follow these steps carefully:

1. Cover the subfloor with 15-pound asphalt-saturated felt at right angles to the direction the floor will be laid.
2. Location and straight alignment of the first course is important. Place a strip of flooring ¾-inch from the starter wall (or leave as much space as will be covered by base and shoe mold), and groove side toward wall. The gap is needed for expansion space and will be hidden by the

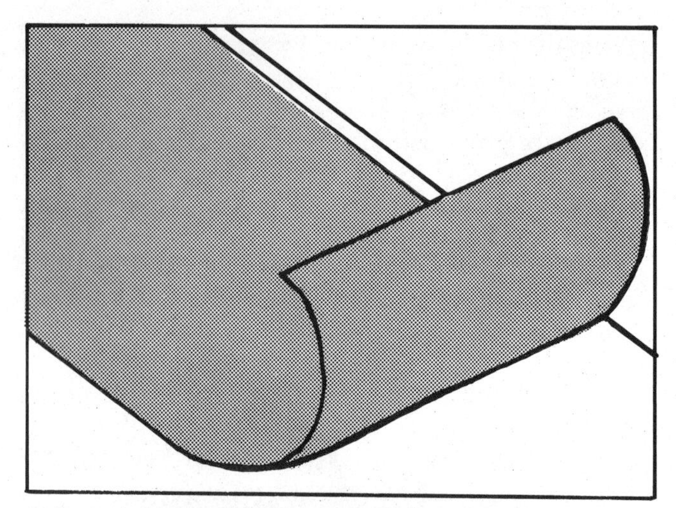

1. Lay asphalt paper over the subfloor.

2. Measure and position first strip.

shoe mold. Mark a point on the subfloor at the edge of the flooring tongue. Do this near both corners of the room. Snap a chalk line between the two points. Nail the first strip with its tongue on this line.

If you are working with screeds on slab, you will not be able to snap a satisfactory chalk line on the loose polyethylene film laid over the screeds. Make the same measurements and stretch a line between nails at the wall edges. Remove line after you nail the starter board in place.

3. Lay the first strip along the starting chalk line, tongue out, and drive 8d finishing nails at one end of the board, near the grooved edge. Drive additional nails at each joist or screed and at midpoints between joists, keeping the starter strip aligned with the chalk line. Predrilling nail holes will prevent splits. The nailheads will be covered by the shoe molding.
4. Nail additional boards in the same way to complete the first course.
5. Lay out seven or eight loose rows of flooring end-to-end in a staggered pattern with end joints at least 6 inches apart. Find or cut pieces to fit within ½ inch of the end wall. Watch your pattern for even distribution of long and short pieces to avoid clusters of short boards.
6. Fit each board snugly, groove-to-tongue, and blind-nail through the tongue according to the schedule shown in the table. Countersink all nails. Take care not to damage the wood with

NAIL SCHEDULE
Tongued-and-grooved flooring must be blind nailed

Flooring Nominal Size in Inches	Size of Fasteners	Spacing of Fasteners
¾ × 1½	2″ machine-driven fasteners; 7d or 8d screw or cut nail	10-12″ apart*
¾ × 2¼	2″ machine-driven fasteners; 7d or 8d screw or cut nail	
¾ × 3¼	2″ machine-driven fasteners; 7d or 8d screw or cut nail	
¾ × 3 to 8 plank	2″ machine-driven fasteners; 7d or 8d screw or cut nail	8″ apart, into and between joists
	Following flooring must be laid on a subfloor	
½ × 1½	1½" machine-driven fastener; 5d screw, cut steel or wire casing nail	10″ apart
½ × 2		
⅜ × 1½	1¼″ machine-driven fastener, or 4d bright wire casing nail	8" apart
	Square-edge flooring as follows, face-nailed through top face	
5⁄16 × 1½	1″, 15-gauge fully barbed flooring brad	
5⁄16 × 2	2 nails every 7 inches	
5⁄16 × 1⅓	1″, 15-gauge fully barbed flooring brad 1 nail every 5 inches on alternate sides of strip	

**If subfloor is ½-inch plywood, fasten into each joist, with additional fastening between joists.*

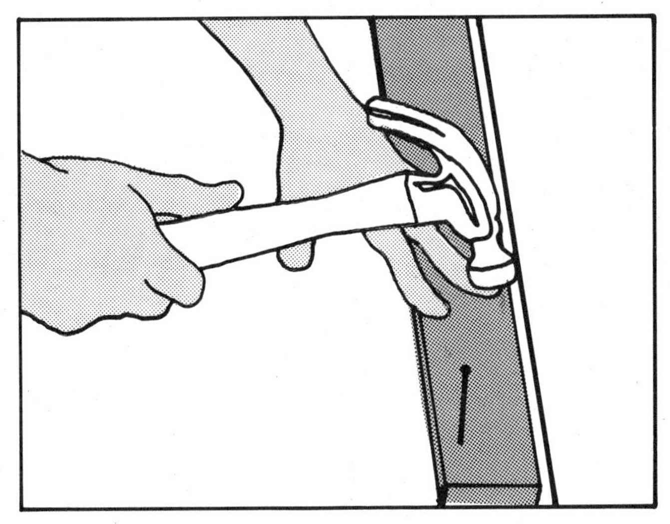

3. Nail first strip into position.

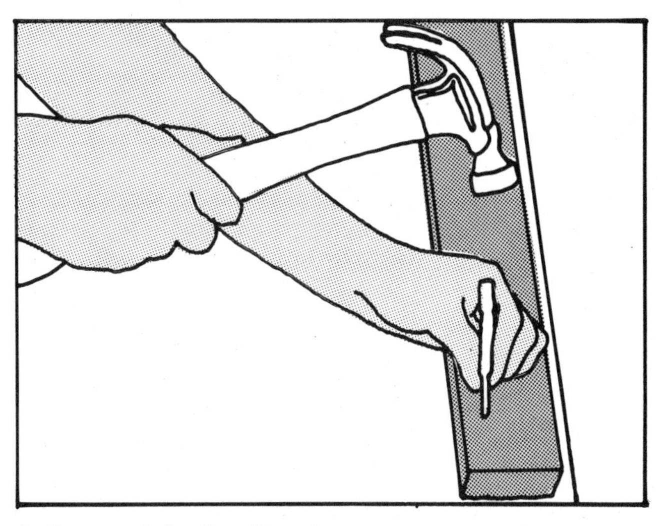

4. Countersink all nailheads.

5. Use a power nailer for easier installation.

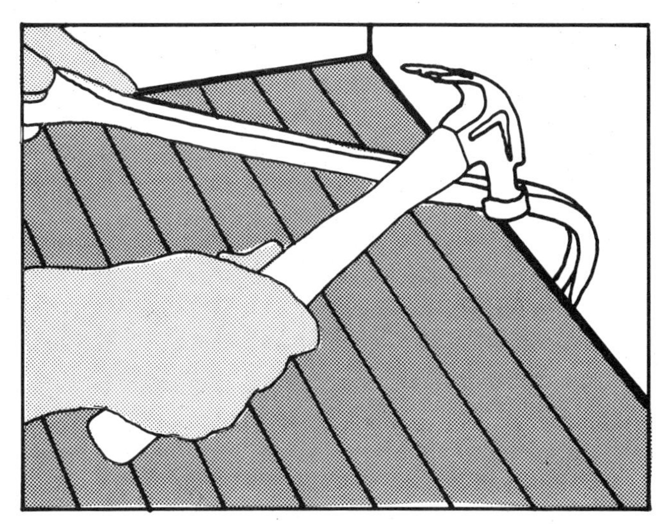

6. Pull last strip tight with crowbar.

hammer dents. After the second or third course is in place, you can change from hammer to a power nailer which is easier to use, does a much better job, and does not require countersinking. The power nailer drives special barbed fasteners at the proper angle through the tongue of the flooring.

When using the power nailer to fasten ¾-inch-thick strip or plank flooring to plywood laid on a slab, be sure to use a 1¾-inch cleat, not the usual 2-inch cleat which may come out the back of the plywood and prevent nails from countersinking properly. In all other applications the 2-inch cleat is preferred. When nailing directly to screeds (no solid subfloor), nail at all screed intersections and to both screeds where a strip passes over a lapped screed joint. Nail across entire face of board. Since flooring ends are tongued and grooved, end joints of adjacent strips should not break over the same void between screeds.

7. Continue across the room, ending on the far wall with the same ¾-inch space allowed on the beginning wall. It may be necessary to rip a strip to fit. The final few courses will again have to be blind-nailed. Use a crowbar to pull boards tight. Face-nail the last two or three courses.
8. Avoid nailing into a subfloor joint. If the subfloor is at right angles to the finish floor, do not let ends of the finish floor meet over a subfloor joint.
9. Nail shoe molding to the baseboard, not to the flooring, after the entire floor is in place.
10. Fill all nail holes with an appropriately colored putty.

If you have square-edged plank flooring, you will have to attach the boards to the subfloor with screws. Even some tongue-and-groove planks require screws for added strength and decoration. Many manufacturers provide planks with predrilled countersunk holes for the screws. If your particular plank flooring does not have predrilled holes, you will have to do this yourself.

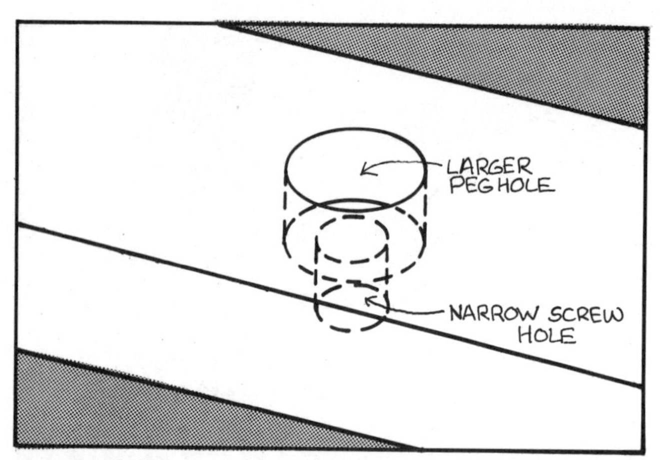

For plank floors with wooden pegs, countersink screw in narrow hole, then fit peg into larger hole to conceal the screwhead.

1. Depending on the width of the plank, countersink two screws at the end of each plank and at intervals along the plank. Do not use too many.
2. Use screws that are the proper size: 1-inch screws if the flooring is laid over ¾-inch plywood, 1¼-inch screws if laid over wood joists or screeds.
3. Glue wood plugs into holes over screwheads.
4. Some manufacturers also recommend blind-nailing in addition to using screws.

Caring for Wood Floors

Wood floors, properly installed and finished, are one of the easiest floorcovering surfaces to keep clean and new looking year after year. Only minimum care is required, usually just a weekly vacuuming or dust mopping. Occasional buffing will remove scuff marks. Wax only once or twice a year or as often as needed in high-traffic areas.

When caring for and maintaining wood floors, remember that water is an enemy of wood. Never wash or wet mop wood floors. Even if the finish on the wood floor is impervious to water, water can seep between the boards or tiles, leaving dark stains or causing the boards or tiles to warp.

No matter what type of floor-care products you use on your wood floor, always read and follow the manufacturer's instructions to make sure the product is compatible with the type of floor you have.

There are two principal types of finishes used on wood floors—penetrating seals and surface finishes. Each requires about the same care; but when it comes to removing stains or restoring the finish in heavy-traffic areas, methods vary. So it helps to know what type of finish was used on your floors.

As a general rule, you can be sure your floor was prefinished at the factory if it has vee-shaped grooves along the edges where the boards join and sometimes where the ends butt. Unfinished plank flooring may also have grooved edges, but usually this is a mark of a prefinished floor. Once you have established the floor as prefinished, the flooring manufacturer can tell you whether or not it has a penetrating-seal finish.

If the floor has no grooves, it was in all likelihood finished by craftsmen after installation. To determine what kind of finish was used, call the builder or floor finisher if possible. When in doubt, it is safest to assume that a surface finish was used. Treating a penetrating-seal finish as though it were a surface finish can do no harm; whereas a surface finish treated like a penetrating seal is likely to be ground away.

Penetrating Seals This is the finish recommended for most residential floors. As its name implies, the sealer soaks into the wood pores and hardens to seal the floor against dirt and certain stains.

At the surface, it provides a low-gloss satin finish that wears only as the wood wears. Because of this, color may be added to the liquid sealer at the time of application, and the eventual effects of traffic will be far less apparent than with other finishes that only coat the surface. When an area does begin to show wear, it can be refinished easily. The new application will blend into the old without lap marks or other signs of repair.

The beauty and wear-resistance of wood floors finished with a penetrating seal may be further enhanced by paste wax. A buffed wax coating forms a barrier against the most frequent kind of abrasion and can be easily renewed.

A penetrating seal may be used also as an undercoat for surface finishes, serving as a stain to color the wood before the surface finish is applied. The surface finish used should be compatible with the penetrating seal, or it may peel.

Surface Finishes There are four basic types of surface finishes used on wood floors.

- Polyurethane—This blend of synthetic resins, plasticizers, and other film-forming ingredients produces an extremely durable surface that is moisture-resistant. It is the best choice for a kitchen floor and is available both in high-gloss and mat finishes. Some polyurethane manufacturers say no waxing is required, but you will probably get better wear and appearance if you give it the same care as other surface finishes.
- Varnish—Depending on the type of varnish used, the finish will be high, medium, or low in gloss. Varnish tends to darken with age and is difficult to touch up. It dries slowly. If the quality is good, varnish will provide a highly durable surface; if not, it tends to become brittle, to powder, and to show white scars.
- Shellac—This is a popular finish for floors in houses built in certain areas of the country. It dries so fast that two coats can be applied in one day and the floor used eight hours later. Liquid

spills can cause hard-to-remove spots on a shellac finish, and the abrasive action of footsteps creates frictional heat that softens the finish and permits entry of dirt. Waxing is essential to protect the finish.

- Lacquer—Even faster drying than shellac, lacquer requires real skill in application or brush marks will show. It produces a tough, high sheen; but this sheen is difficult to maintain, and scuff marks show easily.

Removing Stains

Most stains can be prevented or minimized by keeping the floors properly waxed and by wiping up any spills immediately. When removing a stain, always begin at the outer edge and work toward the middle to prevent it from spreading.

Waxing

If your floors are new or newly refinished with either a penetrating sealer or a surface finish, apply a liquid buffing wax/cleaner or a coating of paste wax. The wax will form a protective barrier for the finish to keep out dirt and potential stain-causing matter so that your floors will stay beautiful and resist wear.

Liquid buffing wax is easier to use than paste wax and is selected more often. For that reason, liquid is recommended, with these two cautions: (1) use a wax that is designated for hardwood floors and (2) do not use a liquid that has a water base. Check the label; only a solvent-base product should be used. Solvent-base waxes will have the odor of dry cleaning fluid.

Follow the manufacturer's directions for applying the wax. Buff it well, preferably with a 12-inch ma-

Stain Removal Chart

Stain	Removal
Dried milk or food	Rub spot with damp cloth. Rub dry and re-wax.
Water	Rub spot with No. 00 steel wool and re-wax. If this fails, sand lightly with fine sandpaper. Clean spot and surrounding area using No. 1 steel wool and mineral spirits. Let floor dry. Apply matching finish on floor, feathering out into surrounding area. Wax after finish dries thoroughly.
Dark spots; ink	Clean spot and surrounding area with No. 1 steel wool and a floor cleaner or mineral spirits. Thoroughly wash spotted area with household vinegar. Allow it to remain for 3 or 4 minutes. If spot remains, sand with fine sandpaper, feathering out 3 to 4 inches into surrounding area, re-wax and polish. If repeated applications of vinegar do not remove spot, apply oxalic acid solution directly on the spot. Proportions are 1 ounce oxalic acid to 1 quart water or fractions thereof. Oxalic acid is a bleaching agent. After it is used, the treated floor area will probably have to be stained and refinished to match the original color.
Heel marks, caster marks	Rub vigorously with fine steel wool and floor cleaner. Wipe dry and polish.
Animal and diaper stains	Spots that are not too old may sometimes be removed in the same manner as other dark spots. If spots resist cleaning efforts, the affected flooring can be refinished.
Mold	Mold or mildew is a surface condition caused by damp, stagnant air. After seeing that proper ventilation is provided for the room, the mold can usually be removed with a good cleaning fluid.
Chewing gum, crayon, candle wax	Apply ice until the deposit is brittle enough to crumble off. Cleaning fluid poured around the area (not on it) soaks under the deposit and loosens it.
Cigarette burns	If not too deep, steel wool will often remove them. Moisten steel wool with soap and water to increase effectiveness.
Alcohol	Rub with liquid or paste wax, silver polish, boiled linseed oil, or cloth barely dampened in ammonia. Re-wax affected area.
Oil and grease	Rub with kitchen soap having a high lye content, or saturate cotton with hydrogen peroxide and place over stain; then, saturate a second layer of cotton with ammonia and place over the first. Repeat until the stain is removed.
Wax buildup	Oak floors that have not had proper care may acquire wax buildup. Strip all the old wax away with mineral spirits or naptha. Use cloths and fine steel wool and remove all the residue before applying new wax. It is a good idea to perform this complete stripping job every now and then instead of using the liquid cleaner/wax process. Stripping removes all the old wax and dirt which build up inevitably over a period of time and partially hide the beauty and color of the wood grain.

chine buffer available from rental companies. You can buff small areas by hand with clean cloths.

Routine Care Vacuuming is the best way to remove surface dust and dirt before it gets ground into the wax and dulls its luster. Vacuuming also pulls accumulated dust from the grooves of prefinished and plank floors. Dust mopping occasionally is also an effective way to control dust and dirt. When floor luster has dulled a bit and scuff marks begin to show, you can often restore the original beauty by simply machine or hand buffing.

Special Care After four to six months of wear, inspect your floors closely to see if there has been a dirt buildup or if the wax has discolored. If your floors were originally finished in a dark tone, the finish may be lighter in traffic areas. If this is not apparent, just apply a new coat of wax over the old and buff it well to restore luster. If such conditions do exist, use a combination liquid cleaner/wax. For dark floors, choose a buffing wax in a compatible dark color. Spread it with a cloth or fine steel wool. Rub gently to remove grime and the old wax, then wipe clean. Let dry 20 minutes or so and buff. If dull spots remain after buffing, apply a second coat and repeat.

Repairing the Finish

Small areas of floors finished with penetrating seal can be repaired successfully without professional help. With special care and skill, you may also be able to repair varnish and polyurethane finishes yourself. Such repair may be necessary after stain removal or water damage. Use steel wool to smooth out the affected boards and an inch or two of the surrounding area. Then brush on one or more thin coats of finish, feathering it into the old finish to prevent lap marks. Allow plenty of drying time between coats and then wax well.

Do not attempt this if you have a lacquer or shellac finish since these are almost impossible to patch successfully. For a small, relatively inconspicuous area, you might get by with steel wool cleaning followed by paste wax. You will not get an exact match, but this may serve as temporary repair. The alternate is sanding to the bare wood over the entire floor and applying new finish.

Extensive damages to a wood floor may require replacement of a board or tile. Use the following steps to replace strip or plank flooring boards:

1. Drill several holes across the width of the board at several points along the board's length. Drill only to the underlayment.
2. Split flooring between holes with chisel. Make as many splits as possible. Remove excess wood.
3. Remove nails and clean space.
4. Measure and mark replacement strip.
5. Remove lower half of groove side with chisel. Plane the edge smooth.
6. Tap replacement strip into position.
7. Drill pilot holes for nails, spacing them 12 to 15 inches.
8. Nail carefully. Countersink nailheads and fill holes with putty.
9. Finish to match surrounding floor.

Refinishing Wood floors that have become unsightly from years of wear or neglect can be restored to their original beauty. Machine sanding removes the old finish and exposes new wood. With the application of a finishing material, floors are like new again. While the highly skilled home craftsman may want to undertake the task of refinishing, it is usually advisable to have a professional floor refinisher do the work to be assured of best results.

Paint should never be used on a wood floor. It will not stand up to traffic, and it hides the natural wood characteristics.

Cracks and Squeaks

All the wood in your home will contract or expand according to the moisture in the air. Doors and windows may swell and stick during rainy seasons. In dry, cold weather, cracks and fine lines of separation may appear in wall cabinets and furniture.

The same reaction to humidity or the lack of it is happening constantly in your wood floors. Tiny cracks between edges of boards may appear when unusually dry conditions are produced by the heating system. This can usually be corrected simply by installing a humidifier. With even regulation of moisture in the house, the floors will benefit.

When interiors become damp in rainy weather, boards may expand so that edges rub together, causing a squeak. Improper fastening of the floor or subfloor can also cause squeaks. To correct this, first try lubrication. A liberal amount of liquid wax may do the job. Or sift a small amount of powdered soapstone, talcum powder, or powdered graphite between adjacent boards where the noise occurs. Another method is to drive triangular glazier points between the strips, using a putty knife to set them below the surface.

If that does not work, drive 2-inch finishing nails through pilot holes drilled into the face of the flooring. Nails should go through both edges of the boards. Set them with a nail punch and hide with matching color putty and wax.

The best solution requires more work and can be accomplished only where there is access from beneath the floor (crawl space or basement). This involves placing wood screws from below. They are inserted through the subfloor and into the finish floor to pull the finish floor down tight.

Carpeting

Carpet manufacturing has experienced incredible advancement during the past several decades. The types, colors, and patterns of carpets available today are too numerous to list. Discussion in this chapter will be limited to the most popular styles and fibers.

Carpeting originated in central and western Asia where it was painstakingly made by hand of sheep's wool colored with natural dyes. Carpets were originally used to cover earth floors, but as weavers developed their skills in carpet making, silk, silver, and gold threads were used with wool yarn. As carpets became more ornate, they were more frequently used as an interior design element in the homes of the wealthy. Today some of these same beautiful carpets can be seen hanging in art museums.

Natural wool carpeting was the standard of quality for hundreds of years, but manufacturing methods and technology advanced and now carpeting of synthetic fibers compares very favorably with wool carpeting. Although most carpets are no longer constructed entirely of wool, carpeting still serves the same two basic functions it always did: a practical floorcovering material and an interior design element. Both the functional and aesthetic aspects should be considered when selecting the proper carpet for your home. A carpet must offer performance as well as beauty.

Carpeting provides a colorful, warm, soft, and quiet surface. Rich colors are often woven into luxurious textures that are appealing to the eye and comfortable underfoot. Carpeting can be used to unify the decor of several rooms or enhance the interior design of a single room. Since carpeting is the background for activity in a room, it must set the desired tone and mood for a room.

Carpeting does have a few drawbacks as well. It does not have the permanence of tile or wood floors, and carpeting generally requires more care and maintenance than the other floorcoverings.

Unlimited Selection

Because carpeting is constructed in many different ways and is available in a wide variety of fibers, textures, and colors, actual selection can be difficult without a basic awareness of carpeting in general. The better informed you are, the more quality you can purchase within your budget; and you will be assured of buying the right carpet for the right area.

Buying quality carpeting is the key. Quality carpeting, which offers excellent performance in addition to beauty, is more expensive but will last longer. With proper care and maintenance, quality carpeting will last from 12 to 16 years or longer. There are several factors that identify quality which will be discussed in the following paragraphs. These factors are construction, yarn twist, fiber, backing, color, and cost.

Construction

Because carpet construction is so varied, selection can be trying. Construction determines a carpet's texture—the way it looks, feels and wears. There are many types of carpeting ranging from soft and elegant to durable and easy to maintain. The type of construction you select should be determined by the amount of traffic the carpet will have to withstand.

Over 95 percent of carpeting today is manufactured by a method called tufting. Tufted carpeting is made by high-speed machines that stitch hundreds of yarn-threaded needles through a primary backing fabric to form loops or tufts. An adhesive coating is applied to the primary backing to hold the tufts in place. A secondary backing is then applied. The loops may be left uncut or may be cut totally or in selective areas. Some of the carpet constructions made by tufting include:

- Level loop pile carpeting—this is the most basic type of carpet. All loops are of approximately the same height. Level loop pile offers great wearability but is limited in surface texture effects and appearance. It is an excellent carpet for high-traffic areas.
- Multilevel loop pile carpeting—is very similar to level loop pile except that loop height varies which provides a more interesting sculptured texture. It too offers good wearability and durability.
- Cut pile carpeting—is basically a level loop pile in which each loop is cut on the tufting machine into two tufts. It has a soft texture called plush. There are, however, several other types of plushes offering different degrees of texture, depth, and soil and wear resistance. These variations result from the amount of twist applied to the yarn (discussed in the next section). The two most common plushes are:

  A. Velvet plush—has tightly packed individual tufts that form a texture resembling the softness and smoothness of velvet. The yarn of this prestigious type of carpeting has little twist, so the tufts blend together into a dense pile that creates interesting shadow effects on the surface.

  B. Saxony plush—has more twist and is deeper and denser than ordinary cut pile plushes. Un-

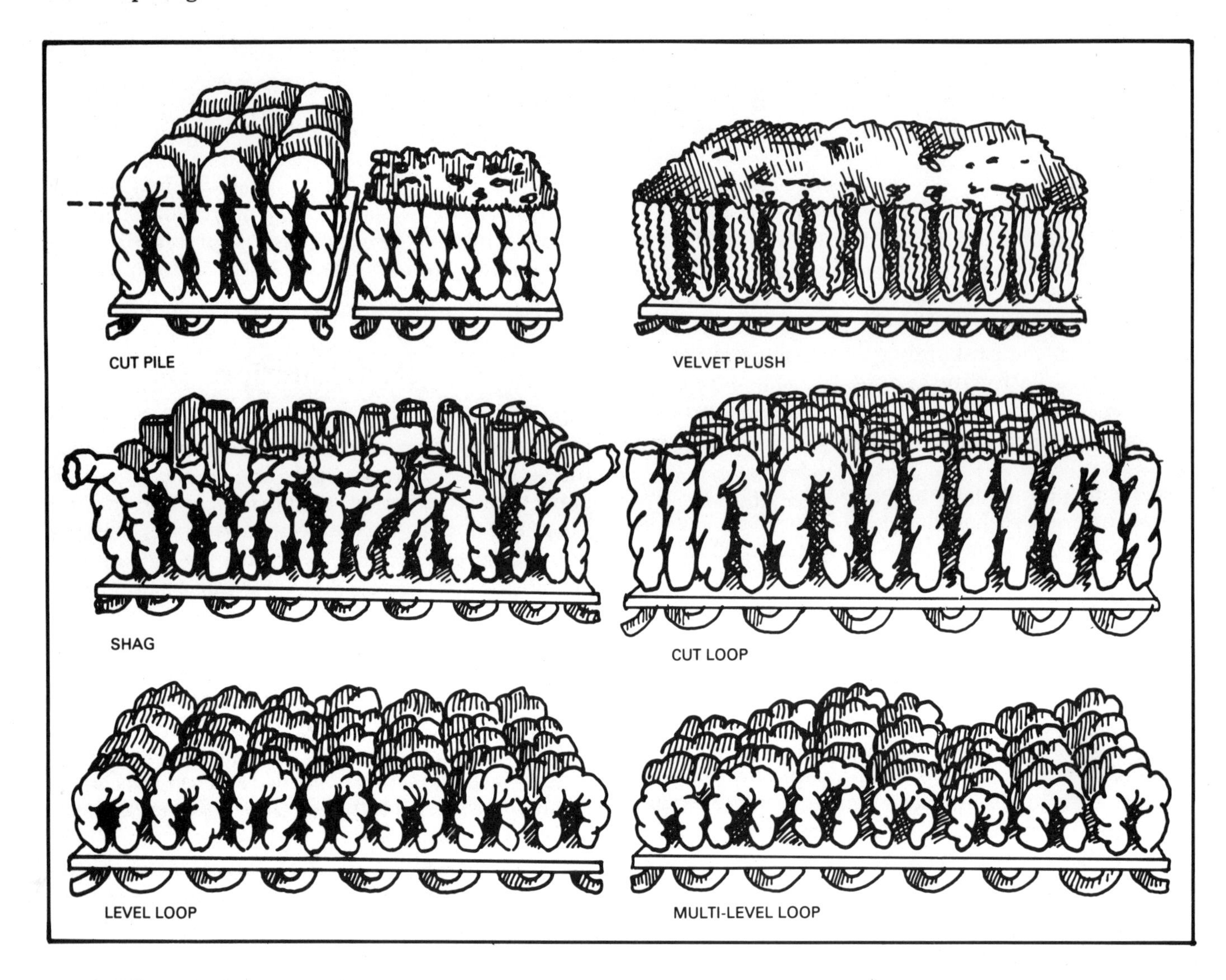

like the blending of velvet plush tufts, the individual tufts of Saxony plush are distinguishable.

- Cut/loop pile carpeting—combines cut piles and either level loop or multilevel loop carpeting to create stunning textural effects. Only some of the tuft loops are cut on the tufting machine. The purpose of this type of construction is to combine the higher performance benefits of loop carpet with the softer texture of cut pile carpeting.
- Shag carpeting—has longer, more widely spaced tufts so they lie in random directions, giving the carpeting an informal shaggy appearance. It is available in several lengths and is mainly used in casual settings.

Another type of construction is the woven carpet. The surface yarns and backing material are intertwined on large looms. This process is slower than a tufting machine, so woven carpets are generally more expensive. There are several types of weaves, each with its unique technique for incorporating the surface yarns into the backing fabric. Each technique produces different effects. The most popular weave is the velvet weave. A velvet loom creates a carpet similar to the loop carpeting created by a tufting machine. In fact, many of the same styles and intricate textural designs are available—cut pile, cut loop, level loop, mutilevel loop, and plush.

The other common type of construction is needlepunching. In needlepunching, thousands of carpet fibers are placed randomly on a fiber mesh. When the fibers reach a specific height, thousands of needles on a needle-bonding machine are punched through. The punching action compresses the fibers and backing into a feltlike solid mass. Additional layers can be added by repeating the process. This type of carpeting is relatively inexpensive, and the styles are somewhat limited.

Consider the following suggestions when selecting the carpet construction to meet your needs:

- The more rows of yarn per inch the carpeting has, the greater its density. Density varies greatly, and, generally, the greater the density, the better the quality. Dense carpeting wears better, resists soil and stains better, and keeps its original shape and appearance better. Bend a carpet backward.

The less backing you see, the denser the carpet. Denser carpets are usually more expensive.

- Look for good balance between yarn rows and stitches. Carpeting with high pile feels good underfoot and is usually more expensive; however, manufacturers of cheap carpeting sometimes achieve high pile by reducing the number of yarn rows or stitches.
- Look closely at the weight of the yarn used in the carpet. Different carpet fibers have different weights, so compare yarn weights of carpets made of the same fibers.
- Purchase roll carpeting in widths wide enough to fit the room in order to eliminate seaming. Seaming is difficult and time-consuming.
- Use jute-backed carpeting in low-traffic areas and rubber-backed carpeting in high-traffic areas.

Yarn Twisting All carpet yarns are twisted prior to manufacturing. The more turns per inch, the less bulk in the carpet and the better the carpet's performance. To determine twist level, look closely at the cut tips of an individual tuft. They should be neat and well defined, not frayed and spread apart. When you must decide between pile density and twist level in carpeting for high-traffic areas, twist level is more important. Tightly twisted yarn springs back and is less likely to unravel or mat. The tighter the twist, the smaller the yarn diameter; so more yarn can be tufted per square yard.

Carpet Fibers Choosing a fiber today is somewhat easier than it was in the past. For several years carpeting manufacturers attempted to create a variety of carpet qualities by using several synthetic materials in many combinations. But today with the technological advances made in the carpet industry, only five basic fibers are commonly used.

- Wool—is the main natural fiber used to any extent today. It makes a luxurious carpet with good performance characteristics and holds dye particularly well. Wool is not plentiful, however, so wool carpets are generally the most expensive to produce. Wool is not as durable as the synthetics.
- Nylon—over 80 percent of the carpeting produced today is made with nylon. Economical to produce, its performance features are far superior to most other fibers. It is the strongest synthetic fiber, offering excellent wearability and abrasion resistance. It is also soil and moisture-resistant.
- Acrylic—was introduced in the 1950s. It has the same soft, warm, luxurious feel and appearance as wool and offers even better resiliency and cleanability. Acrylic carpets are more expensive to produce than nylon carpets, so their use has diminished in recent years.
- Polyester—also resembles wool quite closely. Polyester carpeting offers high bulk that imparts resistance to matting and abrasion. Polyester is not as colorfast or as resilient as nylon.
- Olefin (polypropylene)—is one of the newest synthetic fibers available. It is strong, resists soil and stains, and resists fading. Most indoor-outdoor carpets are made of this material. This type of fiber is also commonly used in backing.

Synthetics are generally easier than wool to handle and install.

Backing The material to which the individual fibers are attached is called the backing. There are two basic styles of backing: conventional and cushion-backed. Yarn fibers are attached by tufting or weaving to a primary backing of jute (natural), polypropylene (synthetic), or foam. The backing is then coated with a latex or thermoplastic adhesive material to permanently anchor the yarn fibers to the backing. Carpeting left in this state is called conventional. When a secondary backing of foam rubber, sponge rubber, vinyl, or polyurethane is secured to the primary backing with an adhesive, the carpet is called cushion-backed. Conventional carpeting is usually installed over a foam rubber pad to improve resiliency and wearability. It must be installed under tension with stretchers and tackless wood strips. Cushion-backed carpeting is easier to install because it can be glued or taped down. A pad is not necessary. Cushion-backed carpeting is less expensive than conventional but tends not to last as long.

Color Most people let color govern their carpeting choice when, in fact, it is one of the least important considerations. Nevertheless, several factors should be considered when choosing color. Very light carpeting shows dirt and stains quicker, and extremely dark colors show dust and dirt. Carpeting generally covers a large area, usually an entire room surface; the color of a carpet will often dominate a room. Carpet color must, therefore, blend with the surrounding furnishings and decor. A carpet depends on a dye or dyes for color. Make sure the dyes used for the carpet you plan to purchase will not fade under continual exposure to light or after cleaning with strong carpet-cleaning detergents. Various types of lighting can cause a color to appear differently. Daylight is the best light source for judging a carpet's true color. Take samples of wallcovering, furniture fabrics, paint, and drapery material to the store with you when choosing a carpet color to make sure it will harmonize with each material. Better yet, if carpet samples are available for you to look at in your home, you will be able to see how the color works in the place you intend to install the carpet.

Cost Purchase the highest quality carpeting possible within your price range. Quality materials which generally have a higher initial cost than those of lesser quality, will prove more economical in the long run.

Where to Use Carpeting

Carpeting can be used virtually anywhere in the home as long as you purchase carpeting suitable to the function of the room. Spills and grease are common kitchen and dining room problems. Moisture can be a problem in the bathroom. Hall and entryways are subjected to heavy foot-traffic and plenty of dirt. Carpeting in the living room must be beautiful and yet easy to clean. Bedroom carpeting does not have to be nearly as durable as the carpeting used in any of the areas mentioned previously. There is a carpet for each of these areas which has the physical characteristics to match the demands of the area. Choose a carpet that best suits your lifestyle.

Living Room The most beautiful, luxurious carpet that you can afford should go here. A multilevel loop or cut pile nylon is a good choice since this will help minimize maintenance, particularly if the room receives relatively hard daily use or food and beverages are occasionally served in the room. If the living room is more of a showplace with little actual functional use, a velvet or Saxony plush will improve the room's appearance. If you can afford wool, this is the area to install it.

Family Room Carpeting used in this area must be durable and easy to maintain. A tightly twisted cut pile nylon is probably most appropriate for the wear and tear this room would receive. If you have pets, avoid loop carpeting.

Halls, Stairways, and Entryways These are traditionally the heavy-traffic areas in most homes. Choose a dense low pile nylon carpet. Multicolor carpeting tends to hide dirt. Installation on stairs is tricky, so professional help may be necessary.

Bedrooms The bedroom generally receives less traffic than any other area in the home. Here comfort,

Shag carpeting creates a casual, comfortable floor for this contemporary living room.

beauty, and warmth are the important features to be gained with the selection of a carpet. A luxurious, light-colored carpet may be most appropriate. Children's bedrooms and playrooms require low pile and easy maintenance carpeting.

Kitchens and Dining Rooms Carpeting in these areas has many demands. It must be wear-resistant, stain and soil-resistant, long-wearing, and easy to maintain. Tightly twisted treated nylon fibers of low or medium height should be used.

Bathrooms The essential factor here is moisture-resistance. All construction materials, including fibers, primary backing adhesives, and secondary backing, must be moistureproof; otherwise bacteria, mold, and mildew will ruin the carpeting in short order.

Preparing the Subfloor

One of the primary features of carpeting is that very little work is required to prepare a proper subfloor. There are two basic carpet styles, conventional and cushion-backed, which are installed in different ways. Conventional carpeting is generally applied with tackless strips, and cushion-backed carpeting with adhesive. Both types of carpeting, however, can be laid over most dry, level, clean floors or subfloors. Cushion-backed carpeting that will be applied with adhesive generally requires greater care in subfloor preparation.

Clear the entire working area of all furniture, furnishings, or any other loose objects such as heating/cooling registers. Remove baseboards and molding carefully to avoid damaging them. Number each piece for accurate repositioning. Clean the area thoroughly.

Wood Subfloor

Wood subfloors, whether plywood panels, wood boards, plywood on concrete, or plywood nailed to screeds, are suitable for carpet installation provided they are clean, dry, smooth, and in good condition structurally. They should also be free of all oil, grease, wax, and dirt. Fill all joints or cracks wider then 1/16 inch with a high quality latex patching compound. Sand when thoroughly dry. If you plan to use adhesive, read and follow the manufacturer's directions for subfloor preparation.

1. Check and repair, if necessary, the joist framework as described in the basic floor construction chapter. Nail down all loose panels or boards. Covering a wood board subfloor with an underlayment of 1/4-inch plywood can strengthen the subfloor substantially.
2. Remove any grease or oil stains with a commercial chemical cleaner. Remove paint or finish by sanding, particularly for cushion-backed carpet.
3. Vacuum the surface thoroughly.

Concrete Subfloor

Concrete makes an acceptable subfloor only if it is moisture-free. To test concrete for moisture content, refer to the subfloor preparation section of the wood flooring chapter. Use any of the tests described to determine the moisture content of the concrete you plan to install carpeting over. If the concrete subfloor is dry, install a vapor barrier of polyethylene film or asphalt paper for on-grade or below-grade surfaces.

1. Test the concrete for moisture.
2. Sand or grind all high spots and ridges level.
3. Patch all cracks wider than 1/8 inch with latex patching compound.
4. Remove all grease and oil stains. Sand or remove paint from the surface. Curing compounds are often used to help concrete dry. Such compounds are usually incompatible with most carpet adhesives, so they must be removed from the surface.
5. Sweep and vacuum the area.
6. Many carpeting manufacturers recommend sealing a concrete floor with a sealer that is compatible with the adhesive you plan to use.

Existing Floorcoverings

Carpet can be installed over several existing floorcovering materials. The floor must be clean, dry, and structurally sound. Floorcovering materials must be securely bonded and in good condition. All waxes, paint, and varnish must be removed.

- Resilient floorcoverings—Make sure the resilient sheet or tile floorcovering is firmly adhered to the subfloor. The adhesive residue is probably not compatible with the carpet adhesive if you plan to glue down the carpeting. Remove adhesive residue with a commercial solvent. Ask your floorcovering dealer to recommend a suitable product. Clean the subfloor thoroughly. If the existing resilient surface is in good condition, remove any wax and clean thoroughly.
- Wood strip, plank, or parquet flooring—If there is no detectable movement of strip or plank flooring boards, the carpet can be installed directly over them. If you will glue down the carpeting, paint and finish will first have to be removed. A layer of 1/4-inch underlayment can be nailed over the existing wood floor if board movement is noticeable or if you want to avoid stripping the finish.
- Ceramic tile floors—Securely bonded ceramic tile floors make an excellent base for carpet installation. Fill all grout lines with a latex-based flashing compound and allow to dry. Clean surface thoroughly.

Installing Carpeting

Carpeting is installed in three basic ways: by stretching and fastening to tackless strips nailed around the perimeter of the area, by setting in adhesive applied to the subfloor, or by installing carpet squares as you would any type of tile whether wood, ceramic, or resilient. Conventional carpeting with jute or synthetic backing is generally applied using tackless strips. Cushion-backed carpeting and carpet squares are installed with adhesive. In some cases cushion-backed carpeting can be installed with double-faced carpet tape.

Carpeting is difficult to install. Most residential installations are done by professionals for this reason. If, however, you are patient and confident of your skills, the following step-by-step instructions will enable you to install most types of carpeting.

Make sure the subfloor is prepared properly for the type of carpeting you plan to install. Take the scale drawing of the area to be carpeted with you to the dealer in order that the exact amount of carpeting needed can be determined. Store the carpeting at room temperature, so it will be properly conditioned to the area at the time of installation.

Tools

Most of the tools required for tackless strip carpet installation are common household tools. A few specialized tools, however, are required to ensure and simplify proper installation. It is not economical to purchase these tools as most carpeting dealers will rent or loan specialized tools. If not, many tool rental businesses carry these tools. Assemble all tools prior to installation.

- General purpose tools—claw hammer or tack hammer, nail set, razor knife or carpet knife, chisel, staple gun and staples, nails, putty knife, and heavy-duty shears.
- Measuring tools—tape measure and straightedge, chalk and chalk line.
- Specialized tools—knee-kicker, power stretcher, seaming iron, row-running knife, seaming kit, and tackless strips.

Tackless Strip Installation

This is the most common installation method today. Wood strips, with rows of projecting pins angled to the wall, are glued or nailed to the floor around the perimeter of the area to be carpeted. The carpeting is then stretched to the strips where the pins firmly grip the edge of the carpeting. The excess is then trimmed off along the wall.

The secret to successful carpet installation using this method is proper seaming and even, uniform stretching. These are also the most difficult steps of the procedure for most do-it-yourselfers.

1. Install tackless strips around the entire perimeter of the room. Tacks should be angled toward the wall. Leave ¼-inch space between strip and wall.

Nailing the Tackless Strips

Many carpeting dealers have tackless strips available, or you can make your own by cutting thin strips of wood from ¼-inch lightweight plywood. Stagger many sharp ¼-inch tacks at a 60-degree angle from the face of the strip. Strips are attached to the subfloor with nails, masonry nails, or glue. Strips must be installed at every point where the carpeting will be stretched and secured.

1. Place lengths of tackless strips around the entire perimeter of the area to be carpeted, with tacks angled toward the wall. Ends of the strips should butt one another.
2. Nail the strips into position, allowing about ¼ inch between the strip and the wall. This space should be consistent. Use a tack hammer to pound nails. A heavier claw hammer must be used to drive masonry nails into a concrete subfloor. Always wear safety goggles. Nails cannot be used on a ceramic tile subfloor. The strips should be fastened with high quality contact cement. Follow the manufacturer's directions.
3. Use binding bars at door openings.

Laying the Padding

Jute and synthetic-backed carpeting require foam padding for increased resiliency, protection, and comfort. Most paddings have a wafflelike imprint which should always be positioned up. Select a padding that

2. *Roll out the padding, overlapping the tackless strips.*

3. *Staple padding to the subfloor near tackless strips.*

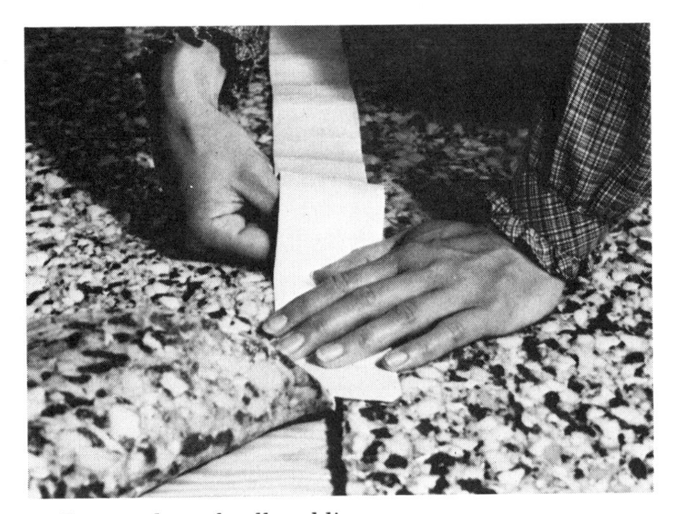

4. *Tape and staple all padding seams.*

5. *Trim excess padding.*

is best suited to the type of carpeting you are buying.

1. Roll out the padding so it overlaps all tackless strips. It is usually necessary to cut several pieces to cover the entire working area.
2. The padding is fastened to the subfloor either with adhesive if the subfloor is concrete or ceramic tile, or with staples for wood or resilient subfloors. With a staple gun, staple the padding to the subfloor along the inside edge of the tackless strips. Space the staples 4 to 6 inches.
3. Tape all seams together. Staple both edges to the floor.
4. Trim the excess padding with a razor knife along the inside edge of the strips. The tackless strips should not be covered with padding.

Cutting the Carpeting

Unless the area you are carpeting is relatively small, you will probably have to cut the carpeting to fit the area. Use the scale drawing you made of the area for accurate room measurements.

1. Roll out the carpeting and, using a flexible tape measure, carefully measure for the cuts you have to make. Allow for a 3-inch overlap around the edges of the room and for any necessary seams. If you have a patterned carpet, make sure the patterns will match.
2. Carpeting must be laid so that the pile lay of adjoining sections faces the same direction. If the direction of the pile lay is laid in one direction for one section of carpeting and in a different direction for an adjacent section, the color and shading of the two sections will be different.
3. Cut-pile carpeting is cut from the back. Loop-pile carpeting is cut from the front. For cut-pile carpeting, notch the front of the carpeting to indicate where the cut is to be made. Turn the carpeting over and snap a chalk line between the two notches. Cut only the backing with a carpet or utility knife. For loop-pile carpeting,

6. Cut carpeting, allowing 3-inches overlap on all sides.

snap a chalk line on the carpeting and using a row-running knife, cut the carpet backing along one row if possible.

4. If there are obstacles, such as posts, lay the carpeting up to the obstruction and cut it to fit. The cuts do not have to be perfect because excess can be trimmed later.

Seaming the Carpet

After the carpeting has been cut, chances are you will have to seam sections together. Seams can be made in several ways depending on manufacturer's recommendations. Make sure the edges butt perfectly. Some seaming methods are more effective than others depending on the anticipated traffic level of the area. When possible, seams should run toward the main source of outside light for best results. Remember, the pile lay for each section of carpeting should run in the same direction. Seal all seam edges with latex to prevent unraveling.

- Hotmelt Seams—This technique makes the most durable and least noticeable seam. The use of a high-quality seam tape and a special iron equipped with a heat shield and grooved soleplate is strongly recommended. Take a length of hotmelt seam tape and insert it, adhesive-side up, halfway under the carpet edge. Hold up one carpet edge and slip the iron under the other carpet edge and place directly over the tape. The iron should be at a medium heat setting and moved slowly and evenly. Check the tape periodically to be sure that a complete melt of the adhesive is being obtained. Avoid overheating which can damage the carpet and result in weak seams. Press carpet edges into hot adhesive, making sure that edges butt. Allow the seam to completely set.
- Latex Seams—This seam is acceptable only where the carpet will not be subject to heavy foot-traffic. Cut strips of latex seaming tape and spread the recommended adhesive over them. Center the coated strips beneath the two edges. Apply a thin line of adhesive along each edge of the carpet backing, taking care not to get adhesive on any of the pile. Press the carpeting into position, butting the edges tightly. Allow to dry thoroughly.
- Sewn Seams—The edges of carpeting can also be sewn together. Seal the edges with latex to prevent unraveling. Bottom-sew the backings, using No. 18 waxed linen thread. Use 3 or 4 stitches per inch. Reinforce the back with tape.

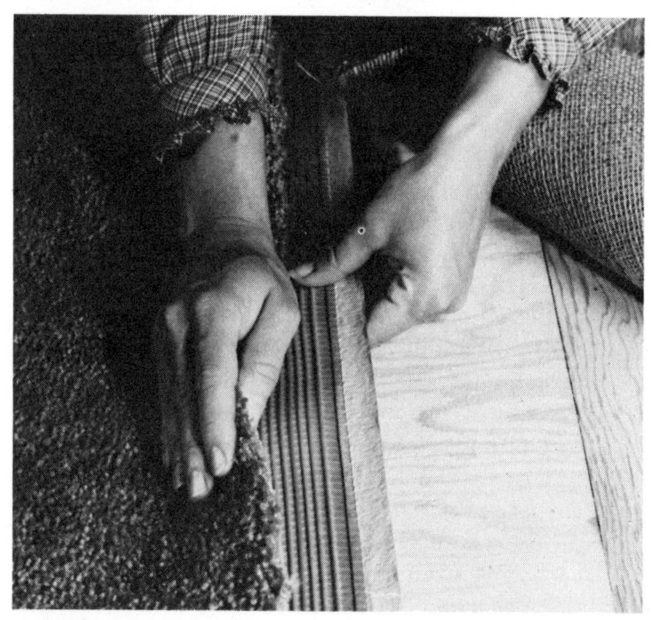

7. Insert hot melt tape under one carpet edge.

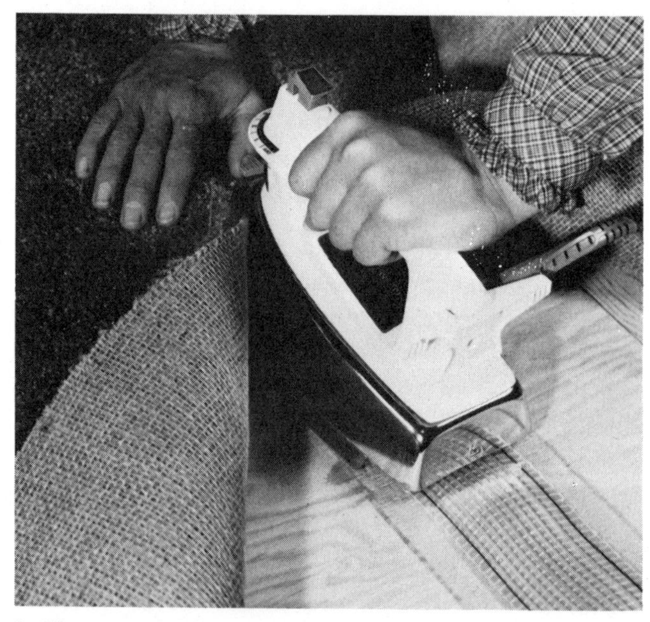

8. Heat tape with seaming iron or old household iron.

9. A knee kicker is used to stretch carpeting.

Stretching the Carpet

When all seams are thoroughly set, the carpet can be stretched into position. The key to proper stretching is uniformity in all directions. Walk over the area to make sure the carpeting lies smoothly on the floor. There should be no air pockets under the carpeting. Use a knife to make small relief cuts at each corner of the room and around all obstacles, so the carpeting will rest flat on the floor. Always stretch carpeting with a slight angle. Do not stretch too tight.

1. Start in one corner of the area. Use the knee-kicker to bump each of the adjacent carpet edges over the tackless strips. Place the head of the knee-kicker an inch from the strip. Bump the pad on the opposite end with your knee while pressing down on the neck with your hand to make sure the teeth on the head are catching the carpet backing. This will move the carpeting up and over the tackless strips, catching the backing on the pins of the strip. Use a slight angle, so the carpeting will be stretched along the wall. Make sure all pins penetrate the carpet backing.
2. Using the power stretcher extended to the proper length (a knee-kicker moves a carpet, whereas a power stretcher actually stretches the carpet), secure the carpeting to the tackless strips at the opposite corner. Stretch the longest edge first. Again angle the stretcher slightly. Do not allow the heel to damage the baseboard or wall. This can be prevented by placing a block behind the heel.
3. When two corners are secured, attach the carpeting between the two corners to the tackless strips.

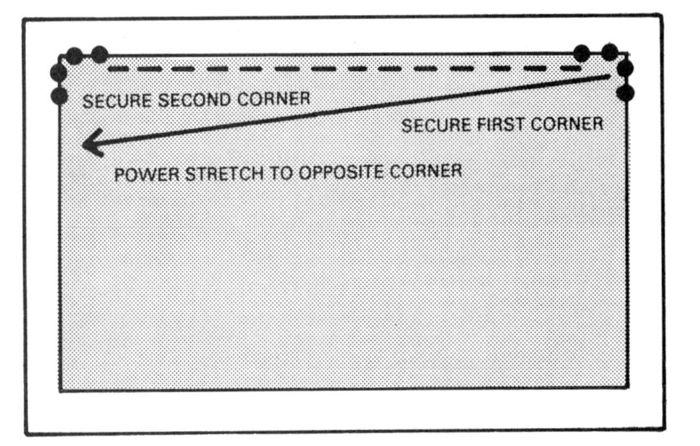

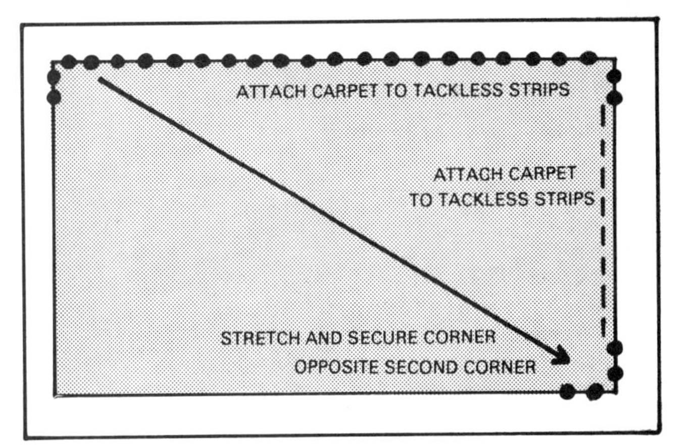

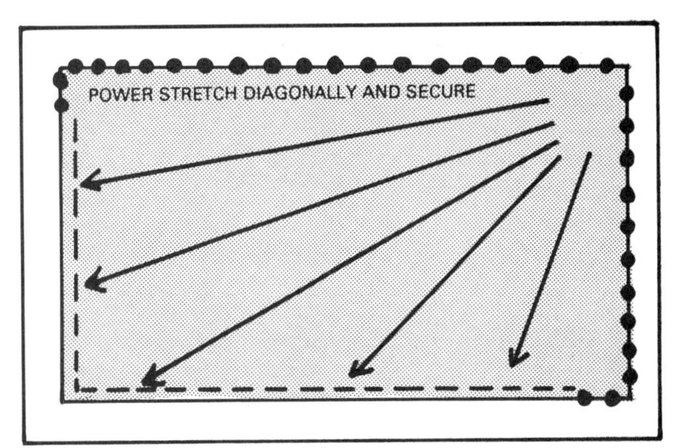

Stretching a carpet properly.

4. Secure the next corner in the same way as described in Step 2.
5. Roll the carpeting over the tackless strips between these two corners. Now two edges of the carpeting are secured to the tackless strips.
6. Starting from the first corner you secured, move the power stretcher along the first edge you secured to stretch the carpet to the opposite wall. Always keep the power stretcher at a slight angle.
7. Starting at the first corner, move the power stretcher along the second edge you secured to secure the final edge. Keep the power stretcher at a slight angle at all times.

Trimming the Carpet

Once the entire carpet is properly stretched and securely attached to the tackless strips, trim excess carpeting between the tackless strip and wall.

1. Cut the carpet with a knife or razor as close to the wall as possible.
2. Use a wide-blade putty knife to force the edge of the carpeting into the narrow gap between the strip and the wall.
3. At the doors, push carpeting into binder bar and, using a wood block as a buffer, pound the top part of the bar with a hammer to bend it tightly over the carpet.

10. Trim carpeting near the wall with knife.

11. Push carpeting into space.

Installing Cushion-Backed Carpet

Cushion-backed carpeting, whether in sheet form or in square tile form, is easier to install than conventional carpeting because stretching is not required and, as the name implies, the padding is bonded directly to the carpet. Cushion-backed carpeting is attached to the floor with adhesive.

Use the adhesive recommended by the carpet manufacturer. Adhesive provides a very strong bond. Once the carpet is down, it cannot be removed without damage to the backing. Double-faced tape is often used in place of adhesive, but the bond is significantly weaker and less permanent.

Carpet tiles are installed exactly like ceramic tile, resilient tile, or wood parquet tiles. Refer to the chapter on ceramic tile installation for proper step-by-step instructions. Just remember to lay all carpet tiles so the pile of all tiles runs in the same direction.

Tools Common household tools are the only tools required to install cushion-backed carpeting. The tools required are: tape measure, chalk and chalk line, carpet or utility knife, scissors, putty knife, and a notched trowel for applying adhesive. Some manufacturers recommend rolling over the carpet with a 150-pound roller. These are available at most rental businesses.

Cutting the Carpet Before rolling out the carpet-

1. Snap chalk line for seam.

ing for cutting, make sure the subfloor is completely clean and properly prepared.

1. Refer to the measurements on the scale drawing of the area you initially made. Measure the carpeting for the first cut, allowing 3 inches overlap for all edges and necessary seams. Pile lay for all sections should run in the same direction. If you are installing a patterned carpet, make sure the patterns match at all seams.
2. Cut the carpeting along measurements as you would conventional carpeting.

Seaming the Carpet Seaming cushion-backed carpeting is relatively easy when compared to seaming conventional carpeting.

1. Snap a chalk line on the subfloor where the seam will be located.
2. Line up the edge of one carpet section with the line.
3. Position the edge of the second section so it overlaps the first edge slightly.
4. Fold back both sections and apply adhesive (according to manufacturer's directions) with the notched trowel approximately 2 feet on both sides of the chalk line.
5. Unfold the first carpet section, aligning the edge with the chalk line. Apply a thin coat of adhesive to the edge of the cushioned backing material.
6. Unfold the second section and butt against the first.
7. Press carpet into place. Allow adhesive to dry.

2. Align carpet edges with chalk line.

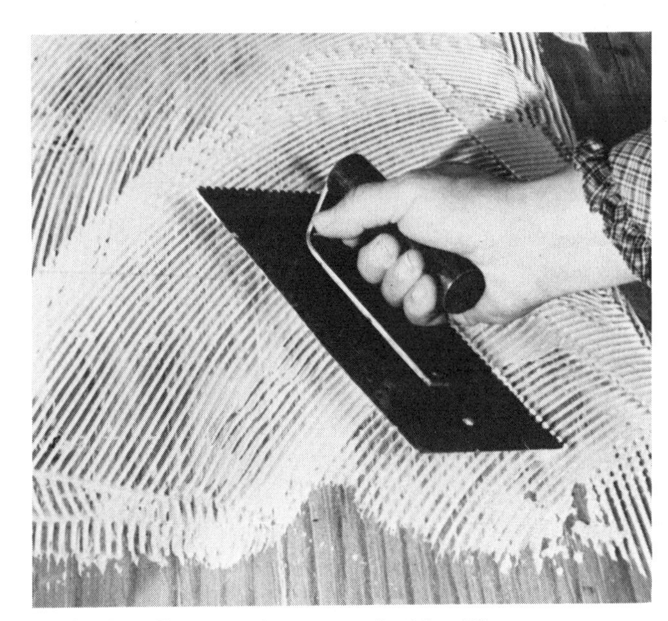

3. Apply adhesive 2 feet on each side of line.

4. Apply adhesive to carpet edge.

5. Butt second section to first.

Applying Adhesive Make sure all seams are thoroughly dry before continuing the installation.

1. Pull back the loose edge of the first section to expose the subfloor.
2. Spread adhesive over the exposed area.
3. Roll the carpeting into position carefully to avoid air bubbles.
4. Repeat this process for the other section.
5. Roll with roller at this point if recommended by the manufacturer.
6. Trim the carpet along the baseboards with a carpet knife.
7. Tuck the edges down against the wall with a putty knife.

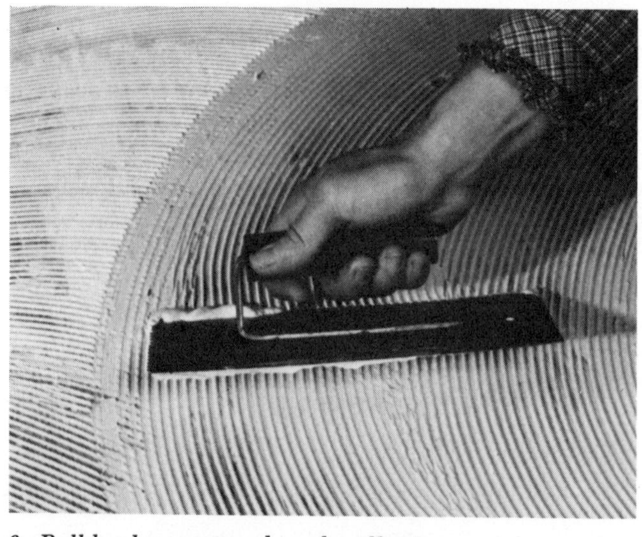

6. Roll back carpet and apply adhesive.

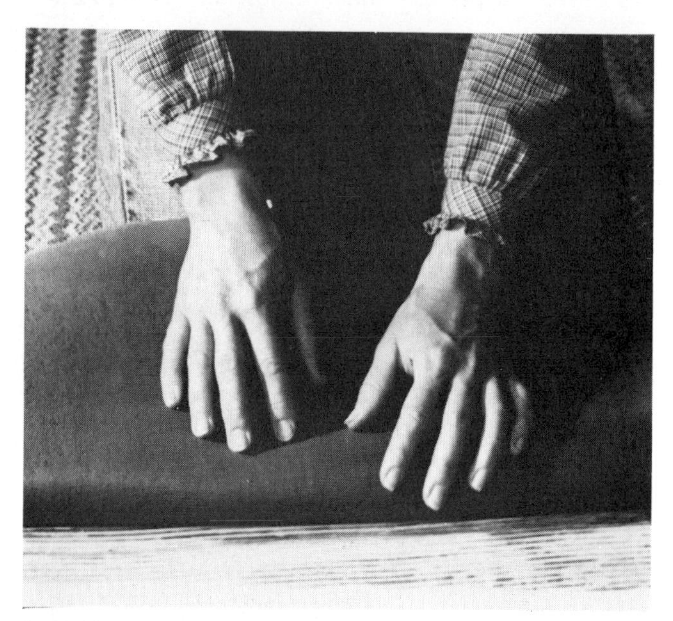

7. Roll out carpet over adhesive carefully.

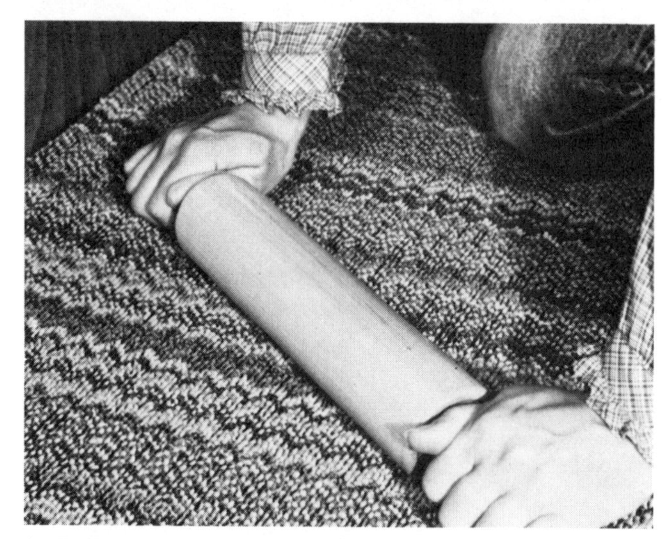

8. Roll carpet with heavy roller.

9. Trim carpet along baseboard.

10. Tuck in carpet edges with putty knife.

Carpet Care and Maintenance

In order to prolong the life of new carpeting and to keep its appearance like new, basic maintenance techniques should be followed regularly.

Preventive Maintenance:

- Use walk-off mats at all entrances to absorb soil and moisture. Clean or replace them as needed.
- Keep adjacent areas free of dirt and substances that may be tracked onto your new carpet.
- Use a good carpet pad, particularly on stairways—it will add considerably to the wear life of your carpet and will give better resilience underfoot.
- Protect your carpet from prolonged periods of direct sunlight with blinds, shades, or awnings.
- Move heavy furniture occasionally to avoid excessive crushing.
- Use furniture rests under furniture legs when possible.
- When moving heavy-wheeled furniture such as pianos or buffets, protect the carpet from damage.
- Do not use chairs or appliances with rollers or casters on carpet with attached foam. Continued use can cause foam delamination.
- Colorful area rugs placed in heavy-traffic areas can enhance your decorating and protect your carpet as well. Remove and clean them regularly; also clean and restore the pile of the carpet underneath.

Regular Care Schedule:

- Establish a regular vacuuming schedule to suit the cleanliness of the air in your community, your family's living habits, the color and fabric you selected, the amount and pattern of traffic in your home.
- Bright, light, and solid colors tend to show soil more readily than multicolors and tweeds and will require more attention. Plush carpets show pile disturbances more readily than low-level-loop constructions. Shags tend to hide soil and dirt; however, vacuuming and raking are needed to maintain the tumbled shag texture. You can expect to vacuum rooms like the family room and areas like the vestibule or hallway more frequently than others due to the amount of traffic exposure.
- At least once a week (more frequently in heavy-traffic areas), give your carpet a thorough vacuuming. If particles of soil and dirt become embedded in the pile, your carpet will lose its bright, new appearance. Even more important, the friction of embedded particles against the sides of the fiber may cause some fibers to split and break.

What to Do for Spills:

1. Take immediate action so that spills do not become hard-to-remove spots or stains. Almost all spills can be satisfactorily removed if given prompt and appropriate attention.
   a. Liquids: absorb as much liquid as possible by blotting (never rubbing) with a clean absorbent material like a tissue, paper towel, or sponge.
   b. Pastes: remove excess by scraping carefully, working from the edge of the spill to the center to prevent spreading. Use a butter knife or a putty knife.
   c. Powders: remove by vacuuming; do not moisten.
2. Blot or gently scrape as much of the substance as possible before it soaks into the carpet.
3. Test shampoos and cleaning fluids on inconspicuous areas of your carpet before applying them to the spot itself. Some cleaning solutions can fade or change the color of certain carpet dyes and make the spot even more unsightly.
4. Avoid overwetting the carpet or allowing dry-cleaning fluid to contact the carpet backing and/or padding.
5. If the area where the spot was removed appears lighter and brighter than the surrounding carpet, the entire area should be shampooed.
6. If neither shampoo nor dry-cleaning fluid will work, try one of the commercial spot-removal kits or contact a professional cleaner.
7. Once the stain is removed, allow the area to dry completely, then brush the pile gently, and vacuum it thoroughly.

Treating Specific Stains:

While it is impossible to anticipate every substance that may be spilled on your carpet, the spot-removal chart in this section is a fairly comprehensive list of spots that may occur in your home.

Do-It-Yourself Carpet Cleaning

1. If you wish to clean your carpet yourself, there are a variety of techniques that you might use. Before starting, however, be sure to remove as much furniture as possible from the room. Use plastic coasters or aluminum foil under the legs of any furniture that remains to protect them from water and to prevent rusting of any metal accessories. Vacuum the carpeting thoroughly, and remove any spots and stains, following the procedures outlined earlier.
2. The following is a general description of four

methods frequently used by do-it-yourselfers. Regardless of which method you use, however, be sure to carefully read and follow the instructions provided by the manufacturer of the cleaning detergent and/or equipment and to select a method that is recommended for the carpet construction.

- Hot-Water Extraction (Steam Cleaning)

Stain Removal Chart

Spot-Removal Chart Instructions:

1. Identify the stain.
2. Follow each step in order by locating the letter shown on the chart in the key.
3. Perform the steps in sequence until the stain is removed. It may not be necessary to perform all of the steps.

	1	2	3	4	5	6	7
Acids	R	A	C	W	T		
Alcoholic Beverages	R	C	V	T			
Ammonia or Alkali	R	V	C	W	T		
Animal Urine	R	W	V	C	W	T	
Ball-Point Pen	X	D					
Beer	R	C	V	T			
Bleach	R	C	V	T			
Blood	R	W	C	A	V	T	
Butter	R	D	C	W	T		
Candle Wax	R	D					
Candy	R	C	V	W	T		
Catsup	R	C	W	T			
Chewing Gum	R	G	D				
Chocolate	R	C	V	Z	D		
Cocktails	R	C	W	T	Z	D	
Coffee	R	C	V	T	D		
Cosmetics	R	D	C	A	V	T	
Cough Syrup	R	C	W	T			
Dye (water)	R	C	A	V	T	P	
Egg	R	C	A	V	T		
Fruit	R	C	A	V	W	T	
Fruit Juices	R	C	A	V	W	T	
Furniture Polish	R	X	D	C	A	V	T
Glue (water)	R	C	A	V	T		
Gravy	R	C	Z	D			
Grease	R	D					
Household Cement (solvent)	R	X	D				
Ice Cream	R	C	A	V	T	Z	D
Inks (water)	R	C	W	T	Z	P	
Inks (solvent)	R	D	X	D	P		
Lipstick	R	D	C	A	V	T	P
Margarine	R	D	C	W	T		
Milk	R	C	A	V	W	T	D
Merthiolate	R	X	D	C	T		
Mud, Dirt, Clay	R	C					
Mustard	R	C	V	T	Z	D	
Nail Polish	R	D	X	D	P		
Oils	R	D	C	A	V	T	
Paint (water base, wet)	R	C	W	T			
Paint (oil base, wet)	R	D					
Paint (dried)	R	X	D	C			
Perfume	R	D	C	V	T		
Rust	P						
Sauces, Salad Dressing	R	D	C	V	T	Z	D
Shoe Polish	R	X	D	C	A	V	T
Soft Drinks	R	C	A	V	T		
Syrup	R	C	V	T			
Tar	D						
Tea	R	C	V	T	D		
Unknown Stains	R	C	Z	D	P		
Vomit	R	C	A	V	T	Z	D
Watercolors	R	D	A	V	W	T	
Wine	R	V	C	T			

Key

A—Ammonia solution (1 tablespoon household ammonia to 1 cup water)—blot.

C—Carpet shampoo solution. Dilute according to manufacturer's directions.

D—Dry-cleaning fluid.

G—"Freeze" the residue using an aerosol chewing gum remover. Shatter with a blunt object, and vacuum up pieces immediately while they are still hard. Note: "freezing" can also be done with ice cubes in a plastic bag.

P—Call professional rug cleaner for advice.

R—Remove excess material (liquids—absorb into clean, white cloth or tissue; solids—scrape lightly; powders—vacuum, do not moisten).

T—Place 1/2-inch layer of white, absorbent material or tissues over damp area under weight for several hours.

V—White vinegar solution (1 tablespoon white vinegar to 1 cup lukewarm water)—blot.

W—Rinse with plain water—blot.

X—Paint, oil, grease remover.

Z—Allow carpet to dry.

Cigarette burns generally must be repaired by replacing the charred area. If the burn is superficial, you may improve the appearance by brushing the surface and carefully snipping away charred tufts with shears.

Method—The hot-water extraction method of carpet cleaning that has become so popular in commercial maintenance is now available to the do-it-yourselfer through the rental system. Portable rental units are readily available and are safe to use on nearly all constructions. Operating instructions are provided with each unit by the manufacturer. Approximately 90 percent of the solution dispensed should be picked up in the vacuuming tank or overwetting of the carpeting will occur.

- Aerosol Method—The shampoo solution foams as it is sprayed directly onto the surface of the carpet from an aerosol container. Cleaning takes place as the foam is worked into the pile with a wet sponge or soft brush. Rinse the sponge or brush occasionally in clean water to remove soil.
- Liquid Shampoo Method—The shampoo solution is applied with either a sponge or an electrical floor-cleaning appliance. The solution is dispersed as a liquid and whipped into a foam, or it can be prefoamed and then applied to the carpet. Cleaning occurs by forcing the foam down into the pile either manually with a brush or sponge or with the electrical appliance.
- Dry-Cleaning Method—This method consists of sprinkling solvent-saturated or detergent-saturated granules onto the carpet and brushing them into the pile, either by hand or with an electrically driven unit designed for this purpose. (Normally this unit can be rented where the granules are sold.) The granules absorb grease and soil and then are removed by vacuuming.

3. Always set the pile in one direction immediately after shampooing and before the carpet can dry. Use a stiff clean brush or a broom. For shags, use a shag rake. Allow the carpet to dry, then vacuum thoroughly.

Additional Precautions:

- Do use a shampoo that dries to a hard, nontacky film which can be easily removed from the pile with vacuuming.
- Don't use a household detergent, soap, ammonia, washing soda, or strong cleaners—they may damage the carpet fiber, and many will leave a sticky film that will cause rapid resoiling.
- Don't use any flammable cleaning agents such as lighter fluid or gasoline.

Performance Characteristics of Your Carpet:

1. Fuzzing or shedding: Loose fiber may work its way to the surface during the first few weeks in your home. Regular vacuuming is all that is necessary to correct this.
2. Pilling: Tiny balls or "pills" of fiber that are attached to the pile by thin strands of fiber may form on the surface of your carpet. Carefully snip the strands with scissors, and vacuum lightly.
3. Sprouting or snagging: If loose ends or "sprouts" extend above the rest of the pile, clip them off even with the pile surface. Never try to pull them out. Then smooth the area with your fingers. Sharp edges on your vacuum cleaner, a child's toy, high heels, or an animal's claw can cause this condition.
4. Crushing or matting: All carpet fibers will crush under heavy loads. Crushed areas where furniture is moved can often be restored by covering the crushed area with a damp, clean white cloth and then applying heat to the cloth with an electric iron, set on a medium-heat setting. Remove the cloth and restore the pile by brushing it lightly. Never apply the electric iron directly to the carpeting.
5. Shading: After your carpet has been subjected to traffic, you may notice areas which appear lighter or darker than other areas. This is particularly true of cut pile carpets like plushes. Do not be alarmed. Shading is a result of the change in direction of the pile due to pressures from footsteps and vacuuming. Brushing the surface may temporarily correct shading; however, shading is characteristic of the carpet, and it should be expected in varying degrees depending on the style of carpet. Shading should not be confused with color fading.
6. Static electricity: The friction of your shoes rubbing on some carpeted surfaces may generate electrical charges that are discharged when you touch a metallic object such as a doorknob. The degree of static buildup varies from fiber type to fiber type. For instance, buildup is usually considerably less in carpets made of polyester, polypropylene, and acrylic than in carpets made of nylon. Static buildup also depends on other factors such as type of shoe sole and amount of moisture in the air. Generally, static electricity lessens as a carpet is trafficked, and it can be minimized by introducing moisture into the air and by treating the carpet with antistatic chemical agents.
7. Shrinkage: Most synthetic fibers used in carpet pile resist shrinkage, rotting, and mildew. Woven and nonwoven jute backing materials, however, will shrink if subjected to excessive moisture. This is why you must be careful to wet only the face of the carpet, not the backing, when shampooing.

Index